ASTRONOMY OF THE ANCIENTS

ASTRONOMY OF THE ANCIENTS

EDITED BY
KENNETH BRECHER
AND
MICHAEL FEIRTAG

The MIT Press
Cambridge, Massachusetts, and London, England

Second printing, December 1981
First MIT Press paperback edition, May 1981

All of the articles in this collection appeared originally in *Technology Review*.

This book was set in VIP Baskerville by DEKR Corporation and printed and bound by The Murray Printing Company in the United States of America.

Library of Congress Cataloging in Publication Data

Main entry under title:
Astronomy of the ancients.
Bibliography: p.
Includes index.
1. Astronomy, Ancient. I. Brecher, Kenneth.
II. Feirtag, Michael.
QB16.A77 520′.93 79-14174
ISBN 0-262-02137-4 (hard)
0-262-52070-2 (paper)

CONTENTS

PHILIP MORRISON

INTRODUCTION

The spin and the orbital motion of the earth, whose great consequences we call *day and night* and *the cycle of the seasons,* are built biochemically deep into the function of each living thing. It goes without saying that they have meant a great deal as well within the symbol systems of our own species, we model builders, ever since we grasped the habit of thought and of language. It is probably more than haphazard that the near-solstice Fourth of July is our glorious Independence Day, while in Canada they mark analogously the First of July.

The night sky is more subtle. The stars and the planets are far from daily life in our busy cities, witnessed more via the little divination games of the astrology columns than through our senses. Even the lively intellectual attention to such profundities as black holes and white dwarfs is due perhaps as much to the magic of words as to the chain of observations by which those concepts are genuinely bound. The name of a star does not usually enter the front pages unless a ship or a horse has also borne it. There is one exception. The indexes show a few references to Canopus, named for the pilot of the Fleet to Troy. But that star also marks the Rudder of the ship Argus. What is this star, bright but not much seen from our latitudes, doing in the occasional story out of the Jet Propulsion Lab or the Cape? Why, it is a modern rudder star, one of the chief targets of the star sensors of spacecraft sent out into orbit, a guide serving to keep the little radio dish oriented toward earth. This is no quaint coincidence. Rather, the real arrangement of the world gave this star its role both for us and for those ancient astronomers, the forerunners of Hipparchus and Ptolemy, who devised the old constellations. For us, Canopus is a bright star, easily picked up, lying at almost right angles to the plane of the ecliptic. Now, if one sensor aims at the sun and another at a fixed direction, the two lines determine a plane in space; and using any plane, the antenna can be aimed aright. There are quite a few other bright stars, but they all lie more or less close to the plane of the planetary orbits. Canopus, combined with the sun, offers the best leverage for accurate pointing; it is indeed a rudder. For exactly the same circumstance, Canopus less than any other bright star partakes of that slow but unceasing drift of the sky with respect to the earth's poles, the precession which over the millenia brought an end to the Golden Age (see Harald Reiche's chapter in this volume). Canopus alone stays fixed, fit rudder for Argus.

This book is one endeavor toward a reconstruction of the past of the human mind, using all the evidence we have, text, myth, and spade, but with a difference. That difference is that in the world of the heavens there are real phenomena, striking or subtle, enduring or transient, which can be invoked today to challenge or to support the inferences of archeologist, epigrapher, historian, or mythologist.

It was just one century ago, in 1879, that an amateur cave explorer, Don Marcelino de Sautuola, and his small daughter found the magnificent painted bison on the ceiling of the cave of Altamira. After years of dispute, the scholars agreed that the bison had been painted by gifted artists at work in Spain two hundred centuries ago, so long that the herds they hunted were unknown in Europe. The net of evidence is strong; today no one doubts that painting and sculpture of a high order are our legacy from the people of the tundra. In the same way, we are now in the midst of a campaign—no doubt full of errors and misconstructions—which looks as though it will establish the origins of a quite sophisticated observational and theoretical astronomy, the birth of an abstract science, if not in the ancient caves, then pretty surely long before the dawn of writing. Signs of it exist in all the world, on the plain of Yucatan with its jungle-covered temples, in the open grasslands of the Canadian prairie, from the scholarly museums of Paris and Athens to the Wilshire hills, in Plato and in the rich Dogon myths, all examples here discussed and analyzed.

Lens, film, and amplifier circuit had not yet come to the aid of the old sky-clerks. Doubtless most of them lacked even the easy elegance of algebra and geometry of the last millenium or two, but one resource they had in plenty, which modern astronomy cannot replace. They had time. Time for supernovas, time for the stars to shift, time for all the unusual arrangements of the planets, for the apparent close approach of Venus, Saturn, Jupiter, time for everything. Our snapshots are more powerful but fleeting. It could possibly be—though I doubt it—that they even had time to see Sirius turn from red to dazzling white, a star evolving without benefit of the "myths of the computer," as Kenneth Brecher puts it. Here, too, there is a pool of wonder; can we learn from the past any more about the universe than we have so far found: the birth of the Crab Nebula, possibly an old dearth of sunspots? Plainly this is a novel and a fascinating road to astrophysics, past the Gorgon's Head, through the whole corpus of classical erudition, to the ethnographic records of peoples the world over. Plainly also, like this volume here in our hands, this enterprise profits from a collaboration among humanist scholars, at home with text and myth, anthropologists, seeking evidence from the handiwork of human beings today and yesterday rather than from written records, and astronomers, with a grasp of what has happened and what might have happened up there in the sky.

For it all goes back to one source: real human beings watching the real sky. There is not likely to be any appeal to ancient astronauts, to mysterious intervenors. It was our forbears, scattered over the continents of the world (and the islands, too) who made the connections real and fanciful, told the tales, piled up the cairns, invented the metaphors. We are all their inheritors, often as unwitting as those who use clocks without a thought for the heavens the clocks model. Indeed, the modern twelve-hour dial is a strange model of the sky, turning twice for each turn of the heavens. The clever digits which today float upon our wrists have abandoned the circle entirely; they still model, but only through the most attenuated and powerful of all metaphors, the natural numbers. But with a little effort we, too, can watch stars trail the sky, the planet Venus cross the sunset horizon, the solstice, and the rising of Sirius slide through the year. If we will do so reflectively, we can share at every glance the roots of the power of human thought, the thought which once married science and literature, art and number, wonder and insight, when thoughtful people were still rather few under this ceaseless sky.

JOHN A. EDDY

MEDICINE WHEELS AND PLAINS INDIAN ASTRONOMY

Among the thousands of organizations active in the United States of America, there is, I am told, an American Society of Witch Doctors, which holds annual meetings at a hotel in one city or another to discuss the problems and practice of that ancient art. Like other professional groups, the witch doctors are concerned about their image, and one of the items that comes up on their agenda year after year is how to get rid of the quacks.

That must be our concern, too, as we take up the question of how much the ancients knew about the sky. Among the sciences, astronomy and archaeology are surely the most speculative, for both are compelled to deal with very incomplete information. It seems obvious that the combination of the two is therefore especially fraught with danger; one must be careful about believing any of it. Yet archaeoastronomy is an exhilarating field, one that is laced with the excitement of controversy and the vigor of cross-disciplinary studies. And I am convinced that in general, the important breakthroughs in science and learning are now to be found in issues that cross the stifling and rigid boundaries of our conventional disciplines.

Is archaeoastronomy a field of quacks? The best known evidence that early man knew and used the sky is surely Stonehenge, in the British Isles. For more than a hundred years, the secret of its alignment with the summer solstice—that is, with the northernmost point at which the sun rises in the course of the year—has been known. Other aspects of the monument's construction are more controversial, such as the claim that it was used to predict lunar eclipses. But Stonehenge does not stand alone; there are at least 900 other structures like it, though not all so grand and megalithic, throughout the British Isles. Many of them have been studied, and by and large their alignments demonstrate an early interest in astronomy. In part by this weight of evidence a basic tenet of prehistory has now been reversed; we must now give up the idea that advanced human knowledge originated only in the Middle East. Evidently Bronze Age astronomy and architecture were fully as advanced in the British Isles as in the Fertile Crescent.

Where else is evidence found? On the coast of France, at Carnac, lines of megaliths run on for kilometers. Astronomical alignments have been claimed for them. In Egypt, the question of astronomical uses of the Pyramids still remains. But I think the evidence is strong that the hidden north passage in the Great Pyramid at Giza is aligned to the lower culmination of the pole star at the time the Pyramids were built. Moreover, the alignments of the pyramidal bases on the cardinal directions were probably laid down by astronomy. And these are but scattered examples.

In concentrating on the Old World, however, we may be examining only the dullest part of the evidence. For probably the clearest case of man's early interest in the skies comes from Central America, where astronomers were priests and priests were astronomers, and the destinies of individuals, of cities, and of nations were thought predetermined by the inexorable clockwork of the sky. Among the Classic Maya, astronomy was not so much an intellectual adventure as a deadly serious game, one that fixed rigid rules for public and private life. It became, in effect, the basis for the highest civilization of the pre-Columbian New World.

As the jungles of Mexico and Guatemala are cleared away, we find temples and pyramids aligned to astronomical phenomena. And we read of astronomy in the stonehewn glyphs and the few codices that escaped burning by the Spanish explorers and clerics. We discover the particular attention paid by the New World's people to the sun, and how the dates of its passage through the zenith determined one of their three sacred calendars—a curious calendar 260 days long. We also read of their attention to that brightest and most beautiful planet, Venus. In the so-called Dresden Codex, a book written by Mayan hands sometime during the first millenium after Christ, we find that some of the strange-looking glyphs represent numbers that encode details of the following celestial cycle:

Begin at a time when Venus has its heliacal rising—that is, when it rises just ahead of the sun and thus disappears rapidly in the brightness of the morning sky. The number of days between that time and the time when it next passes too close to the sun to be seen (the time when again it will rise heliacally) is 236. The number of days until it reappears as an evening star is 90. The number of days until it disappears for a third time, now in the light of the setting sun, is 250. Finally, the number of days until it reappears just before dawn—that is, in a new heliacal rising—is eight.

These numbers, which most of us don't even know today, are all tabulated in the Dresden Codex. Their sum is 584, which is the length of the synodic period of Venus—the time required for that planet to make one apparent trip around the sun. The decoded Dresden Codex thus suggests that the Maya were very close, in about 1000 A.D., to understanding the orbit of Venus, and perhaps to making a map of the solar system. The Dresden Codex, moreover, is one of only three major books that have survived from the Mayan civilization. It is as if a great library had burned down, and only three random documents were saved, one of them containing astronomical ephemerides. What could all the others have contained? Although we shall never know, I like to think that the Mayan civilization may have been about to produce a Ptolemy or a Copernicus—one whom we didn't allow to be born. Michael Coe, the respected Mayanist at Yale University, has bemoaned the fact that Spain sent to the New World, not scientists and savants, but soldiers, who not only overlooked the amazing accomplishments in the New World of learning, but destroyed the civilization that had made them. How sad it is that we didn't give the Mesoamericans another century or two to let them advance their discoveries of the sky and perhaps create, independently, their own physical model of the cosmos. For how exciting it could have been, when the two Worlds met, to compare separate concepts of the universe: that of the Old World, which had long been obsessed with the earth, and that of the New World of America, which had always been more concerned with the sky.

The study of Mesoamerican astronomy has taught us much of what we know about the advanced pre-Columbian cultures of America, which in some respects were fully as advanced as any in Medieval Europe. I suspect that the continuation of this study will produce many surprises. Perhaps it will yield new clues to one of the oldest and most interesting questions of all: How did early man come to the American continents? In how many ways and at how many times? Were there important trans-Pacific or trans-Atlantic contacts? It is known that man did not evolve in the Americas; there are no great apes in the New World from which he could have done so. Accordingly, and from other evidence, it has long been believed that man arrived here rather late, during one of the most recent glaciations, when he walked dry-shod across the Bering Strait and entered the New World through an icy northern door. By possibly 20,000 years ago, he had already reached Patagonia, at the southern tip of the Americas. And by 1,000 years ago, he had achieved a surprising level of culture in Mesoamerica, with an elaborate astronomy, a well developed art, and a written language.

But why, if early man ranged from the northern to the southernmost limits of the New World, did he choose the jungles of Yucatan for his highest achievements? What a curious place for culture to flower! Why did it not blossom in California or Florida, where the living is easy, or in Cambridge, Massachusetts? Why Mexican jungles, of all places? Was there an orderly evolution there? Or might Mexico have been the locus of a significant cultural implant? Archaeologists know that the pottery, masks, toys, and other items made by Olmec and Mayan peoples are much like those made in the Orient at the same time. Moreover, archaeoastronomy is now building a case that the lunar zodiac employed by the Mayans was remarkably like that used by the Chinese and Indian peoples: in both systems, the moon occupies 28 stations in its path around the sky. If a one-to-one correspondence can be established between the two sets of stations, that would seem strong evidence for a late cultural input from the East.

Some of the answers to the teasing questions of how learning spread in America must come from a study of the American peoples adjacent to the flowering cultures of Mesoamerica. These include the American Indians to the north of Mexico: a mélange of many apparently separate peoples, from the relatively sedentary Anasazi to the wandering Plains tribes that roamed through what is now the western United States. In part through archaeoastronomy, we know that long before the time of Columbus, a Mexican influence had spread up into the American Southwest, where we find it in the architecture, the religion, and perhaps the astronomy of the Anasazi and the Hohokam. We also suspect that at one time a Mexican influence crossed the Gulf of Mexico, spread up the Mississippi, and perhaps, late in the sequence of events, led to the building of earthen temple mounds such as we now see at Cahokia near St. Louis. Earthen structures of the Mississippian Mound Builders are remarkably like those of Mexico and probably, therefore, hold astronomical secrets in their remains.

But what of the early Plains Indian? We know less about him, I think, than about any of his contemporaries, even though he was present on the plains west of the Mississippi for at least 20,000 years. We know little, for unlike the Mesoamericans, the Plains Indian had no written language; unlike the Anasazi, the Plains Indian built little; and unlike the woodland tribes of our Eastern Seaboard, whom European man studied rather closely from the time of the first explorations, the Plains Indian was first seen by Coronado in about the year 1540, and then pretty much ignored for almost 200 years. During that time, he had ample opportunity to adapt to what he had learned from the white man, including the use of the horse, which revolutionized his life. We are thus well warned not to think that the early Plains Indians were anything like those we see John Wayne shooting at. For one thing, the early Plains Indians travelled on foot, and their methods of hunting and living must have reflected this.

It is surprising that we know anything at all about early man on the Plains, considering how few of them there were, and over how great an area they were spread. It is commonly said that when Columbus set foot in America there were about as many Indians between the Mississippi and the Rocky Mountains as there are now people who live in my home town of Boulder, Colorado. These 50,000 to 100,000 souls were spread over an area from Texas to Canada, comprising nearly two million square miles. And most of them were always on the move.

Though the Plains Indians built little, they did leave behind perhaps five to six million stone rings, each ranging from five to twenty feet in diameter, made from local fieldstone. The rings often appear in clusters. Because they were so numerous and so common, we can assume they were utilitarian. Most of them are now taken to be the rings left where *tipis* stood, the rocks used instead of stakes to hold down the edges of the heavy hide tents. Archaeologists cautiously date the rings in an approximate way by their diameters, for before the horse the rings were generally smaller.

In addition to stone rings, the Plains Indians built occasional stone alignments and a few effigy figures, such as the figure of a man seen on a field near an eroded gorge in Alberta (figure 1). They also left behind a few stone patterns that are called "medicine wheels." The example seen in figure 2 rests on an eroded river bank in northern Montana. At its center is a small ring of stone, and out from the center radiate spokes which make the wheel look like a drawing of a rayed sun the way your child might draw it in kindergarten. These stone patterns were called medicine wheels because "medicine" suggests the mysterious and the unknown, and the medicine wheels are not fully understood. They are quite unlike the other artifacts that the Plains Indians left behind.

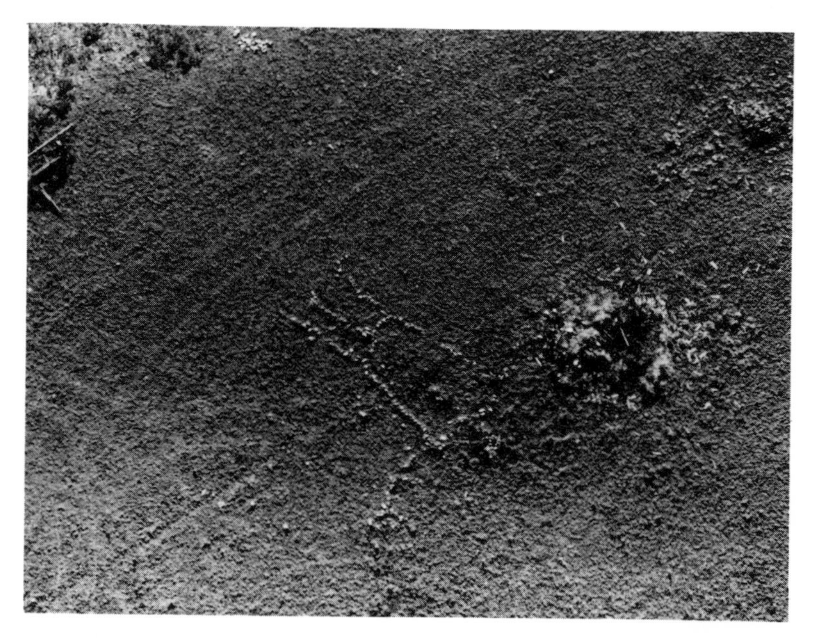

Figure 1
The effigy figure of a man on a short-grass plain in southern Alberta. Though only a few such figures have been found, they appear to have been typical of the Plains Indians. The figures almost always are male; sometimes they are accompanied by a dog. The figure in the illustration is approximately 25 feet long.

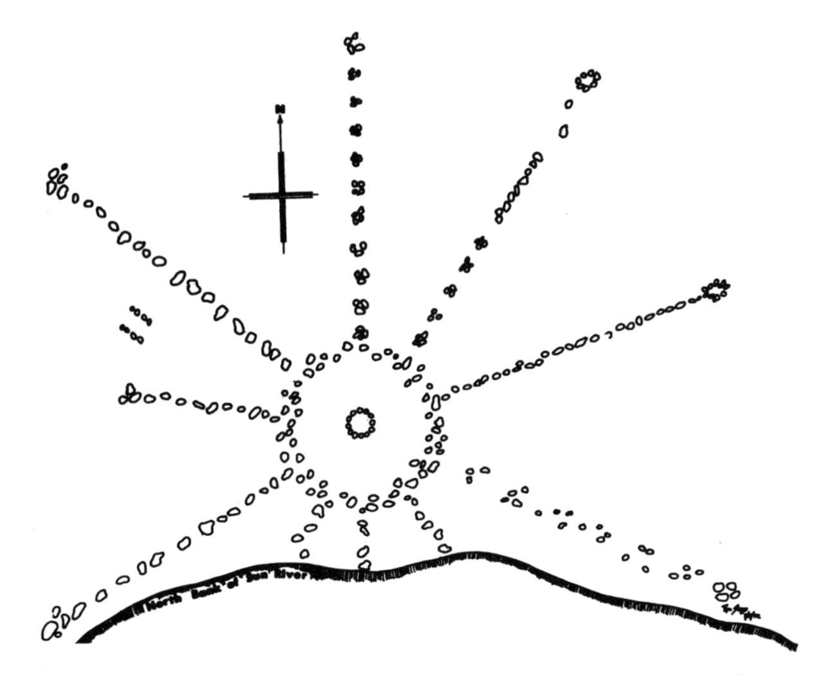

Figure 2
Medicine wheel on the Sun River, west of Great Falls, Montana. Erosion of the river bank has by now destroyed about half of the original structure. Still, the pattern of a medicine wheel is plain: a central ring of stone with spokes radiating outward.

The best known of the medicine wheels (figure 3) is found in the Bighorn Mountains of northern Wyoming, a rich and beautiful land. The Indians loved this isolated range of verdant mountains, and in their futile negotiations with the advancing white culture they tried to hold on to the Bighorns (along with the Black Hills) until the very last.

If you climb to the top of Medicine Mountain, one of the highest of the Bighorn Range, you will follow a narrow road said to be an early Indian trail. It is passable only during a few weeks in summer. Near the summit is the medicine wheel, on a flat shoulder at an altitude of about 10,000 feet. It lies near the brink of a steep precipice; standing there, one can see the lofty peaks that bound Yellowstone National Park, about a hundred miles to the west. The wheel is far from any human habitation, and is protected both by its loneliness and by the fact that its rocks are plain and heavy and don't look worth the

Figure 3
The Bighorn medicine wheel, in Bighorn National Forest, west of Sheridan, Wyoming. The diameter of the wheel is about 90 feet, that of the central hub about 12 feet. Twenty-eight spokes radiate outward from the central hub, and six cairns of stone mark the wheel's periphery. (U.S. Forest Service photograph)

trouble of carrying them off. The Forest Service has also placed a fence around it.

I think that almost every visitor to the medicine wheel is at first disappointed, for it is a crude structure, made from native sandstone. It looks like the sort of thing that a few people could build in an easy day. The wheel includes a central hub—a pile of rocks about 12 feet across and several feet high, hollowed out in the middle. From it radiate 28 spokes. They end at a ring about 90 feet in diameter—not a perfect circle at all, but rather elliptical in shape. At intervals along the periphery of the wheel are six cairns of rock. They are curiously placed: some are on the outer edge of the ring; some are inside the circle; one is entirely outside the ring, at the end of an extended spoke. The peripheral cairns are all hollowed out into U-shaped rock piles. Some of them open one way, some of them another.

The Bighorn medicine wheel was found in the late 1880s by early prospectors. Its existence was soon reported to archaeologists, who went to see it in the early years of this century. They asked the local Indians to tell them what they knew about the wheel. At first, these informants—the Crow, the Sioux, the Cheyenne, the Arapaho, and the Shoshone—said only that they had heard of it, but that they didn't know what it was for, or even where it was. Gradually, though, a few legends came out. The Crow, for example, said that it was there when they came, that it had been built by people who had no iron, or that the sun had built it to show the Indians how to make *tipis.* These stories were sustained somewhat by the Chambers of Commerce in Sheridan and Lovell, Wyoming, who may have wanted to keep the medicine wheel mysterious. I have a clipping from a Casper, Wyoming, newspaper published in 1941 that proclaims: "Solve the riddle of this ancient monument on top of Medicine Mountain and you will solve Wyoming's biggest problem." How simple life was in Wyoming in 1941!

Archaeologists of the 1920s pointed out that the medicine wheel looked much like the plan of a Cheyenne medicine lodge. We are compelled to agree. Figure 4 shows a medicine lodge built by the northern Cheyenne to celebrate the sundance ceremony, a summer ritual that was probably the most common and most important of Plains Indian ceremonies. The lodge includes a central post which might correspond to the central cairn of the Bighorn wheel. Moreover, in this Cheyenne example, there are 28 rafters that radiate outward, much like the 28 spokes of the Bighorn wheel. Finally, in some versions of the medicine lodge there is an altar to the west, where the Bighorn wheel does indeed have a cairn, although somewhat displaced from the true compass direction. These correspondences suggested that the medicine wheel was a replica of the medicine or sundance lodge—perhaps a two-dimensional replica built where wood was scarce above the timberline. It may thus have marked a ceremonial place.

We know, however, that the last refuge of the anthropologist is always in ceremony: if something cannot be explained in any other way, it is usually fairly safe to attribute it to that. I wondered if the medicine wheel could have had a more practical use, a use tied to the sky, which is such a dominant feature of the landscape of the plains. It struck me first that the number of spokes in the medicine lodge and in the Bighorn medicine wheel was close to the number of days in a lunar month. In addition, the positions of two of the Bighorn wheel's cairns, as well as their symmetry about a north-south line, made me wonder if those structures might have been used with the central

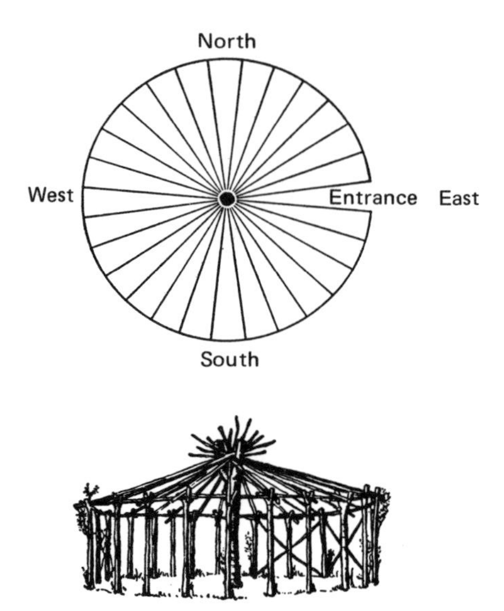

Figure 4
A medicine lodge of the Cheyenne Indians. Such a structure was part of the culture of all Plains tribes; it was constructed specifically for a summer ceremony—the sundance—in a given year, and then abandoned. The heart of the construction was the sacrifice of a tree, whose trunk then became the center pole of the lodge; different tribes had different rituals by which all of this was done. The lodge in the illustration happens to have 28 rafters radiating outward from the center pole; this equals the number of spokes in the Bighorn medicine wheel, but the number of rafters appears to have been variable from lodge to lodge, not just from tribe to tribe. The illustration is from *The Sacred Pipe: Black Elk's Account of the Seven Rites of the Oglala Sioux,* Joseph Epes Brown, ed., copyright 1953 by the University of Oklahoma Press.

cairn to mark the rising of the sun at certain distinctive times of the year. The most easily marked of these is the summer solstice, about June 21, which was, for some, the time of the sundance ceremony. The Bighorn medicine wheel is well placed for a primitive observatory, with good horizons all the way around it.

Several years ago, I set out with my family to test this possibility. In late June, just before solstice, we tried to go to the medicine wheel to watch the sunrise. We failed, for the road was blocked several miles below Medicine Mountain by heavy snow that had fallen the night before. If this were typical of Junes in the past, it seemed unlikely that any Indian would have used the wheel to mark the summer solstice. However, snow comes and goes rapidly in the mountains, and the very next day, on June 20, we were able to reach the site. We found the wheel free of snow; and in fact, considering that mountain peaks are constantly scoured by winds, I suspect that much of the medicine wheel may have been free of snow on the previous day as well. The whisking action of mountaintop winds may have been a reason why the wheel was built there, and in any case the cairns would have poked through any moderate layer of snow.

Our effort to find associations between the sun and the medicine wheel proved to be successful; for standing at the rather distinctive cairn that lies outside the ring, and sighting from its center through the center of the hub-like cairn, as best I could fix those points, I found that I was looking at the distinctive point on the horizon where the sun rises on the morning of the summer solstice—the point where the sun rises farthest north of east in the course of a year. This was confirmed by transit measurements, and by observation of the sunrise in subsequent years. I also found that when I stood at a second cairn and sighted across the central hub, I saw the point where the sun set on the same day. Thus, if the builders of the medicine wheel wanted to mark this distinctive day of the year, they had built in a second way to do so: if the dawn were cloudy they could use the sunset. For on the day of summer solstice the sun sets farthest north of west.

I think it unlikely that the solar alignments involving three cairns of the Bighorn medicine wheel were accidental. However, they leave a number of other cairns to be explained. Were they also astronomical markers? One must be very careful in trying to establish this, for there are many objects in the sky. Given, therefore, a small number of cairns but a great number of stars and planets, not to mention the varying positions of the moon and sun, one is fairly likely to come up with something. I first made calculations to see if the sunrise at winter solstice was marked. It was not, which is not surprising, because the site is accessible only in late spring and early summer. I also checked for markings of the moon's wanderings. They, too, were absent. I then began to look for markings of the stars.

Figure 5 is a time exposure of stars rising before dawn at the medicine wheel (the cairns have been illuminated by flash). The dimmer stars, of course, are the thinner lines in the photograph. Notice that these star trails do not begin until the stars are

Figure 5
Celestial risings before dawn at the Bighorn medicine wheel. The photograph is a time exposure; two cairns of the wheel itself have been illuminated by flash. Note that the trajectories traced by dimmer stars tend to become visible only at fairly great heights in the sky—not at all near the horizon. But one object, brighter than the others, is sufficiently bright so that it becomes visible far lower. That object is the planet Venus—not a star. It happened to rise in alignment with the cairns at the time the photograph was taken, and serves to illustrate the sort of horizon phenomenon to which some of the cairns may have been aligned. (© National Geographic Society; photo by Thomas Hooper)

appreciably above the horizon. The brighter stars, however, can be seen far lower in the sky. I found that one of the brightest stars in the sky rises roughly in line with two of the non-solar cairns of the Bighorn wheel. It is Aldebaran, the brightest star in the constellation Taurus. And from one of the cairns used to observe Aldebaran, two other cairns line up with two other very bright stars. One is Rigel, the brightest star in Orion. The other is Sirius, the brightest star in the sky.

Now all of these stars are situated in the area of the sky in which the sun spends midsummer. It follows that all of them will rise near dawn at the time of the summer solstice; and accordingly it seems reasonable that they might have been noted or used by people who were watching for a summer dawn on top of Medicine Mountain. Moreover, the heliacal rising of one of them, Aldebaran, turns out to be exactly right to mark the summer solstice at the time the wheel was built—determined by archaeologists to be 200 to 400 years ago. It is the only star in the sky that could have served this purpose.

How would the sky have looked if you had stood at the top of Medicine Mountain before dawn on the day of summer solstice several centuries ago? About an hour before dawn, Aldebaran would rise. The pre-dawn sky would already be blue, and all the dim stars would be gone. Indeed, the coming sun would be brightening the sky so rapidly that on this particular day Aldebaran would flash out like a beacon near the horizon, lasting only a matter of minutes before disappearing in the predawn glare. That phenomenon would make this day a distinctive one, for on the previous day Aldebaran would not have been seen at all (the sun's light would have masked it) and on the day after it would have persisted far longer. In short, watching for Aldebaran's flash at dawn would have given a precise indication of the solstice, accurate to within a day or two.

And after that? Each morning after the summer solstice, Aldebaran would climb farther into the sky before the light of dawn would extinguish it. At dawn 28 days after solstice, the bright star Rigel would appear, as Aldebaran had done, but above a second line of cairns. It is interesting that there are 28 spokes in the medicine wheel, but one cannot know whether this connection is fortuitous.

Rigel, like Aldebaran before it, would flash briefly just above the horizon on the first day of its appearance, and on subsequent days would rise higher into the sky before extinction by the light of the morning sun. Twenty-eight days after Rigel's first appearance, Sirius would rise above the third of the cairn alignments that I was able to find. Sirius, too, would flash briefly on the first day of its appearance.

The overall idea, then, is as follows: Aldebaran's brief flashing in the sky—that is, its heliacal rising—would warn you that the day of summer solstice had arrived. An hour later, the location of the sunrise itself would confirm this. So would the sunset that night. One month later, Rigel would appear in the morning sky; and one month after that, Sirius. These latter events may simply have marked off the time during which the mountain could be occupied. The rising of Sirius would have been a good sign to leave Medicine Mountain because winter was coming.

If all this is true, it suggests that the medicine wheel may have been a sacred place, known, perhaps, to a few, who used it as a way to mark off an accurate calendar, or more prosaically, to signal the solstice. That date may have been needed to mark the time of the sundance ceremony. The resemblance of the medicine wheel to a sundance lodge might then be valid, but in the opposite way than was previously supposed: the sundance lodge may have been modeled after a medicine wheel, rather than the reverse.

Could the solar and stellar alignments at Medicine Mountain be accidental? Calculations of the mathematical probability of chance alignment could be given, but I have never believed very much in that approach. It is probably better to examine other medicine wheels to see if any of them show the same alignments. The map shown in figure 6 gives the locations of medicine wheel sites known to me; they are found along the eastern

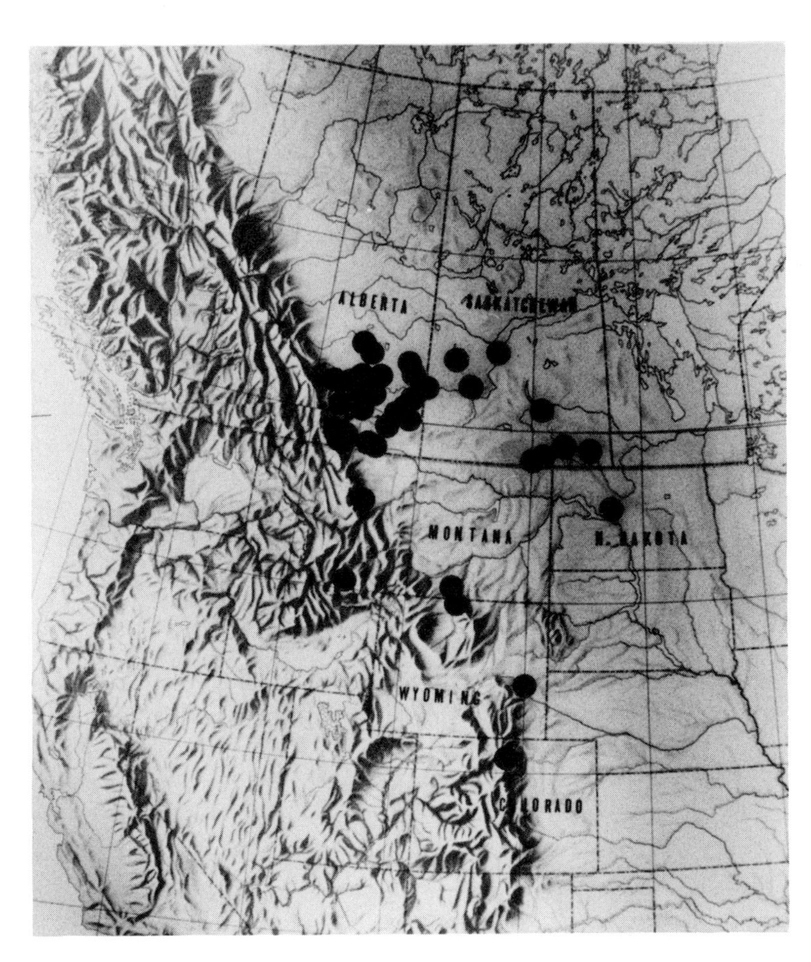

Figure 6
Locations of medicine wheels known to the author. All of them lie at the eastern boundary of the Rocky Mountains or within a few hundred miles farther east.

boundary of the Rocky Mountains, in a broad sweep that stretches several hundred miles to the east over the Great Plains. Most of them lie north of the Bighorn wheel; many lie in Canada.

Figure 7 shows a medicine wheel on a hillside at Fort Smith, Montana, on the present-day Crow Reservation, about 60 miles north of the Bighorn wheel. The Fort Smith wheel is a small circle with five crooked spokes that droop down over a hillside; I suspect that some of the bending is due to soil creep on the hillside, but perhaps they were always bent. The longest of the spokes is aligned to the rising of the sun at summer solstice.

The structure shown in figure 8 may be a primitive medicine wheel in Colorado's Rocky Mountain National Park, at an elevation of about 11,600 feet, near an old trail attributed to Ute Indians. This wheel comprises only a central cairn and two crude spokes. One of them is aligned to the summer solstice sunrise.

Most medicine wheels are found in Canada, in the prairie provinces of Alberta and Saskatchewan. With a grant from the National Geographic Society, and in association with archaeologist Dick Forbis of the University of Calgary and Tom and Alice Kehoe of Milwaukee, I was able to make aerial surveys and ground surveys of about 20 of them. We found that they are very different, one from another; I think they were built at different times, by different people, for different purposes.

Some of the medicine wheels are simple piles of rock. Others are slightly more elaborate: the large central cairn is sometimes surrounded by circles and often by what looks like a gateway. We found that these gateways and other simple features usually point to nothing special in the sky, but rather to something on the ground—namely another medicine wheel, sometimes ten miles away, sometimes 20 or 30. We suspect that some of these simple wheels could have served as landmarks in an otherwise unmarked terrain, for the rolling land around these wheels looks so much like an ocean that it makes one seasick to fly over it. There are few trees or other landmarks.

Still other medicine wheels have very elaborate patterns, much like the Bighorn wheel. The one shown in figure 9, seen from the air in Alberta along the banks of the South Saskatchewan River, is about 200 feet across. It has a central circle that looks like it might have been a *tipi* ring, and spokes that radiate outward and end in small, knoblike cairns.

Figure 7
This and the following figures 8-10 show the variety of structures subsumed under the name "medicine wheel." Here is shown a wheel at Fort Smith, on the Crow Reservation in southern Montana. A small circle and five radiating spokes can be dimly seen; they splay across the lower part of the photograph.

Figure 8
The wheel shown here is a primitive structure, perhaps an early medicine wheel, in Rocky Mountain National Park, Colorado.

Figure 9
This elaborate wheel lies on the bank of the South Saskatchewan River near Medicine Hat, Alberta.

Finally, some medicine wheels are in the shape of effigies. The one shown in figure 10 is in Saskatchewan near Minton, and takes the form of a turtle. The turtle's back is the central cairn of the wheel. Bushes have grown on it, giving the cairn a furry appearance. The body of the turtle includes four legs, a head (with two ears!) and a tail. The Kehoes found that the axis of the turtle points to sunrise at summer solstice.

If we limited our attention to only those wheels that are distinctive, and not just heaps of rock, we almost always found associations with objects in the sky. In this regard, the most interesting site of all is situated atop Moose Mountain, in the rolling prairie of Saskatchewan, southeast of Regina. Figures 11 and 12 show the site. Moose Mountain is really a low ridge, not a mountain at all, and there are no moose living there. I think the ridge resembles the back of a moose from a great distance. At the very end of the ridge, just over the abandoned farmhouse in the photograph, is the highest point, the final bump, on Moose Mountain. It is a place that is associated in Indian legend with the sky: there are stories about an Indian maiden who walked along the ridge of Moose Mountain, fell in love with the sky, and upon reaching the bump was carried away by the sun. The medicine wheel is found at that place. It has a large central

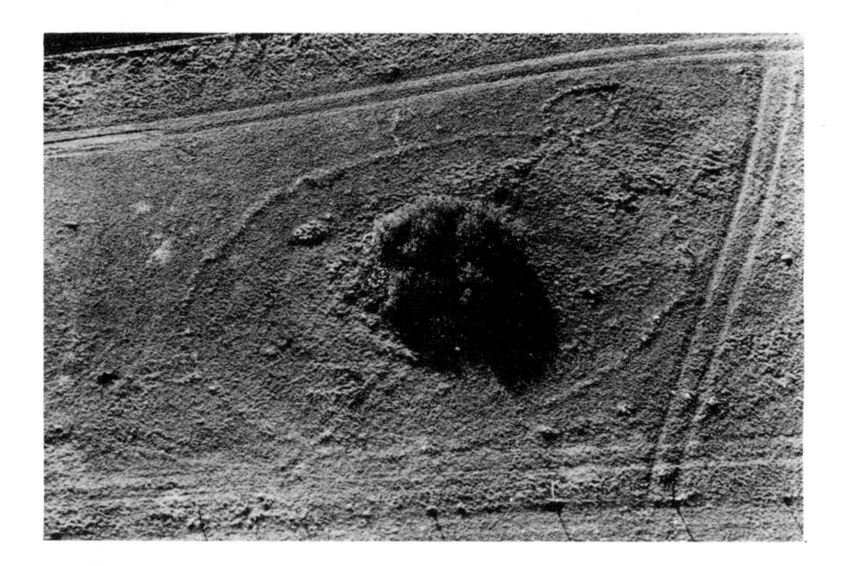

Figure 10
A wheel, found near Minton, Saskatchewan, in the shape of a turtle. Bushes have grown on the central cairn.

Figure 11
The site of the Moose Mountain, southern Saskatchewan, medicine wheel. All of the modest ridge spanning the horizon is the mountain in question—hardly a moose and indeed hardly a mountain. Toward the left of the ridge is a small bump, the site of the wheel itself.

cairn, as do almost all of the Canada wheels. Perhaps the cairn was built up gradually, as passers-by picked up local rock and threw it on the pile. The present total is about 80 tons, all of it picked up nearby. The wheel also has five spokes that give it an overall diameter about twice as large as that of the Bighorn wheel. There is a sixth spoke that is somewhat shorter. All of the spokes are primitive and rudimentary; all have sunk into the ground with age. At the end of each spoke is a cairn.

I have not yet said that recent archaeological excavation of the Bighorn wheel has shown that its central cairn was built first, and the 28 spokes and rim were added later. This made us wonder if the Moose Mountain wheel in Saskatchewan might be an early version of the Bighorn wheel: that is to say, if the Moose Mountain wheel might be the cairns without the later decoration. This seems especially possible when the two wheels are directly compared. Figure 13 does so. On the left is the Bighorn wheel; on the right, the wheel at Moose Mountain. Assuming that the spokes and circumferential ring were late additions at Medicine Mountain, and that the structure first comprised only a central and outlying cairns, the correspond-

Figure 12
An aerial view of the Moose Mountain medicine wheel. Five spokes with cairns at their ends are easily visible, though one is shorter than the others. A sixth cairn is far harder to see. In a sense, however, it is redundant, for it marks the direction to sunset on the day of summer solstice, and a far more prominent spoke marks the direction to sunrise on that day.

ence is very good. For cairn E, which was the sunrise cairn at Medicine Mountain, there is a cairn E at Moose Mountain. For cairn F, there is a similar match, and so on. Not only are the cairns correct in number, they are also all placed at similar relative positions. I think the two wheels could almost have been built from the same set of plans.

We wanted to test this similarity by watching a summer solstice sunrise from the end of the long spoke at Moose Mountain—the spoke corresponding to the solstice cairn at the Bighorn wheel. But the sky was cloudy on that day during our visit; it was impossible to take photographs that would convince anyone that when the sun rose, it did so at the right point. We convinced *ourselves,* however, by some surveying and calculating: the sunrise proved to be off the line of the long spoke by only about half a degree, which is impressive, considering that our uncertainty in fixing the center of the cairns is greater than that.

The other cairns, which at the Bighorn wheel were aligned to three bright stars, proved at Moose Mountain to be aligned to the same three objects: Aldebaran and Rigel and Sirius rose in

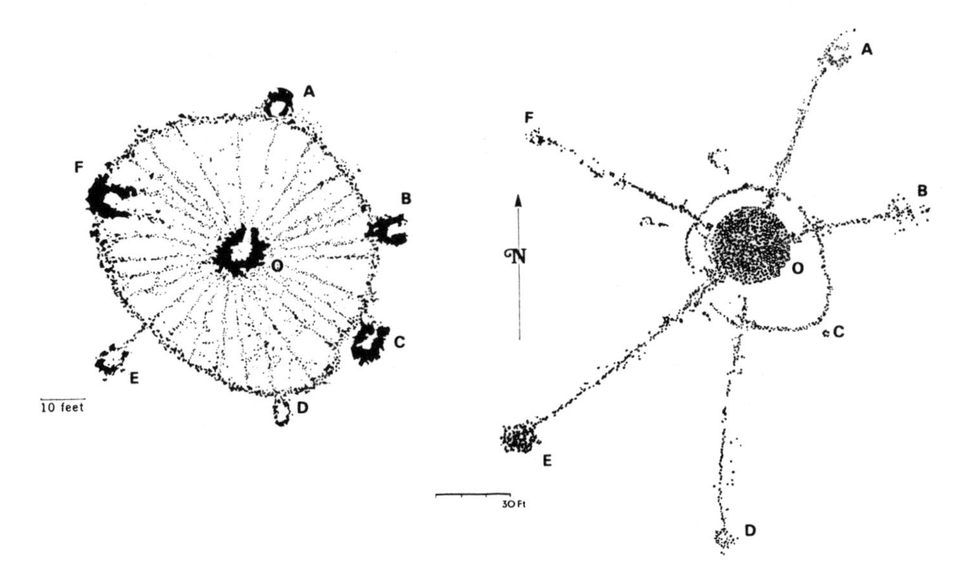

Figure 13
A comparison of the Bighorn (left) and Moose Mountain medicine wheels. The six cairns on each can be placed into one-to-one correspondence, almost as if the same building plan had been used for both.

the fashion described earlier in this article—provided one assumed that the Moose Mountain wheel was built much earlier than the Bighorn wheel. For the stars in the sky precess—that is, they move with the wobble of the earth—and thus the places and times at which they rise change with the years. In the region of the sky that here concerns us, the change has been fairly great. In order for the cairn alignments to match the rising points of Aldebaran, Rigel, and Sirius, I had to hypothesize that the Moose Mountain wheel was built not in 1600 or 1700 A.D., as the Bighorn wheel presumably was, but much earlier than that: roughly at the time of Christ. Figure 14 shows some details of the calculation. One bar shows the approximate of time when Aldebaran rose heliacally: this datum suggests that the Moose Mountain wheel was most likely used between 100 and 500 A.D. The direction of Aldebaran's rise, which is a separate and independent test, gives roughly the same date. Another bar shows the time at which the Sirius alignment would hold true; Sirius moves very slightly, and thus fails to be a good test. Still another bar shows the Rigel alignment, and suggests a time ranging between 200 B.C. and 100 A.D. As for the solstice alignment, the

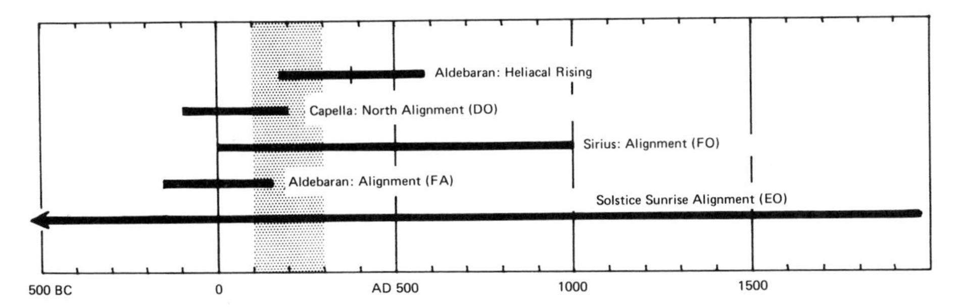

Figure 14
The most probable time of use of the Moose Mountain medicine wheel, based on assumed astronomical alignments. The first bar on the chart shows the span of time during which the rising of the star Aldebaran occurred heliacally—that is to say, with the sun. The second bar shows the time of an astronomical phenomenon not discussed in the text. Consider that the Moose Mountain wheel has a spoke directed northward. The author reasoned that north could be determined by finding a star whose circle around the north celestial pole in the course of a night caused it to touch the horizon; the point of that tangency would be due north of the observer. Such a star, it turned out, was Capella; the second bar thus shows the time during which this tangency occurred each night. The third bar in the chart shows the time during which the rising of the star Sirius occurred in alignment with the spoke marked FO in figure 13. The fourth bar shows the time when Aldebaran rose in alignment with the spoke marked FA. The final bar shows the time when the sun rose in alignment with the spoke marked EO on the day of summer solstice. This last is no test of when the wheel was used: the sun still rises at that direction on solstice day. As the chart's shaded area shows, all of the data combine to suggest that the wheel most probably was used almost 2,000 years ago.

sun moves so slowly that its travels do not help us in estimating a date of use.

Relying upon the data shown in the illustration, I went out on a limb and announced that the astronomical alignments indicated that the Moose Mountain wheel was about 2,000 years old. I don't think that Tom or Alice Kehoe believed me. Nor was I very sure myself.

Then, in January 1977, I got an excited telephone call from Tom Kehoe. In the summer of 1976 the Kehoes had made further studies at the site. Specifically, they had excavated a part of the central cairn, cutting out a wedge so as not to destroy the structure. At the very bottom of the cairn they had found a flat stone floor; they felt sure they were down to the original construction. Under that, they found charcoal; apparently the ground had been burned off before construction started. There was enough charcoal at the site to permit a radiocarbon dating. The result, Tom reported, was that the fire had burned 2,600 years ago, plus or minus 250 years! That seems close enough to support the astronomical dating of the site, and proves that sometimes we astronomers are lucky.

More important, the radiocarbon dating lends credence to the alignments at the Bighorn wheel and other sites. The Indians of the Plains, that people about whom we know so little, seem to have known more about the sky than we may have thought. Yet if we ask Plains Indians today about the sun or stars, or if we examine the oral depositions taken from them during the last century, we find very little, if anything, about practical astronomy. I think it had all disappeared, long before we asked. Practical solstice marking, using sun and stars, had been in Plains lore a long time. For 2,000 years, if we are to believe the Bighorn and Moose Mountain stones, the Plains Indians used the same stars in the same way; yet by the time of the last century it seems to have been all forgotten. With the coming of the white man and his horses and his calendar and his ways, a pure and primitive natural astronomy may have been knowledge no longer needed. How fragile is learning, without the written word!

For further reading:

"Astronomical Alignment of the Bighorn Medicine Wheel," by John A. Eddy, *Science* **184,** 1035–1043 (1974).

"Mystery of the Medicine Wheels," by John A. Eddy, *National Geographic Magazine* **151** (1), 140–146 (1977).

Native American Astronomy, Anthony F. Aveni, editor, University of Texas Press, 1977.

In Search of Ancient Astronomies, Edwin C. Krupp, editor, Doubleday and Co., New York, 1977.

JOHN C. BRANDT

PICTOGRAPHS AND PETROGLYPHS OF THE SOUTHWEST INDIANS

Some years ago, a group of professional astronomers, including myself, became interested in a nebula in the southern sky, the so-called Gum Nebula. It is a large object, presumably a cloud of debris left behind by a violent stellar explosion of the type called a supernova. In fact, it is known that among the debris composing the Gum Neubla is a pulsar, a pulsing source of radiation, believed to be the rotating, extremely dense core of the star that exploded.

We could guess the age of the Gum Nebula because the spin rate of the pulsar is slowly decreasing. The age came out to be eleven thousand years. That places the time the supernova explosion was visible from the southern hemisphere of the earth at roughly 9000 B.C., and recent archaeological work in the southern hemisphere suggests that people lived there long before that time. Now if the supernova that produced the Gum Nebula was as bright as we think supernovas ought to be, and if the exploding star was at the distance from earth calculated for the nebula that now remains, the explosion should have produced a burst of light in the sky fully as bright as the quarter moon. Moreover, the burst should have lasted at least several days, and then grown dimmer over a period of many weeks. Surely even very primitive cultures, faced with two moons in the sky, would have considered this an unusual event. Perhaps somewhere on a stone wall they would have made some record in commemoration. We couldn't know what the record would look like, of course, but we felt we had a scientific duty to perform, so we wrote a paper which was published in *Archaeology,* asking that archaeologists keep the Gum Nebula supernova in mind when conducting research in the southern hemisphere.

We truly believed that discharging our professional obligation in this fashion would be the end of the matter. After all, the journal *Archaeology* was sufficiently obscure for astronomers so that none of our colleagues would snicker at us for our unseemly bit of moonlighting. But in one of those things that happen for curious reasons, *Time* magazine thought that our story was newsworthy. They published a piece on it in March 1972.

Time had no photograph of a "Gum-Nebula pictograph," nor did anyone else. Instead, *Time* used a picture taken by William Miller, who is now retired but used to be the professional photographer at the Hale Observatories. Bill is an amateur archaeologist; the photograph is of a cave painting in northern Arizona, thought to record the supernova explosion of 1054 A.D., the explosion that created the Crab Nebula. Of course, records of a supernova explosion in 1054 A.D. have no direct bearing upon our speculations about records of a supernova explosion that occurred 10,000 years earlier. Still, we thought, the *Time* magazine article won't end our careers.

About four months after the *Time* article appeared, we received a letter from Muriel Kennedy, the wife of the then superintendent of Lava Beds National Monument in northern California. She wrote that the illustration reminded her of a painting in a cave at the monument. I was traveling to California anyway, so I made a sidetrip. The cave painting was marvelous. I was hooked on archaeoastronomy, and have been in the business ever since.

The hypothesis I wish to explore in this article is that western North America (primarily the western United States), contains several images representing the Crab Nebula supernova, visible on earth in 1054 A.D. We know it was visible in that year because Japanese and Chinese literature, specifically the records of the Sung Dynasty, tell us so. Moreover, a comparison of Crab Nebula photographs taken on different dates shows that the nebula is expanding at a rate suggesting an age of about 900 years.

Were the Indians of the Southwest sufficiently astute, and sufficiently interested in celestial phenomena, so that they might have recorded the explosion? I will claim that they were and that they did so. All the evidence is circumstantial, of course, but if the circumstantial evidence is denied, then one must postulate instead a large number of rather strange coincidences.

To begin with, a definition: the archaeoastronomers considering evidence relating to supernovae must in general deal with two kinds of records. The first of these, called a pictograph, is an image made on rock with paint or chalk or a rock that mimics chalk. The second, called a petroglyph, is an image incised in rock by a chisel or an object used as a chisel. It has been suggested that the word pictogram ought to be used for both records, but this idea has not yet caught on.

Now figure 1 shows a petroglyph and a pictograph that Bill Miller first discussed in print in 1955. Both records include a crescent and circle. The circle is often used as a representation of the sun or some other bright object; the obvious candidate for the crescent is the moon. One is therefore encouraged to

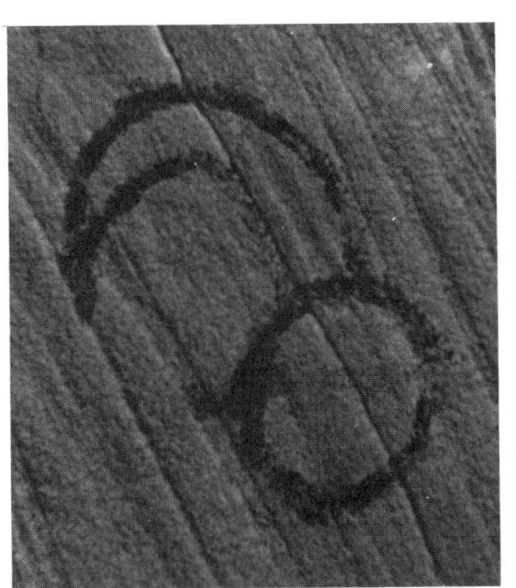

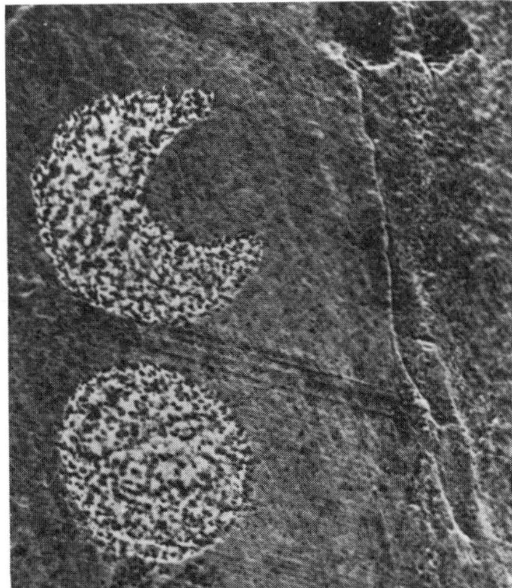

Figure 1
Two examples of rock art thought to represent the supernova of 1054 A.D.—the supernova, or stellar explosion, that created the Crab Nebula. Both of the records shown in the illustration were found in Arizona; both of them are about four to six inches in height; and both of them show a crescent—apparently the moon—in close contiguity to a symbol often used by Indians of the American Southwest to represent a bright object. The record at the left is a pictograph, an image painted or chalked on rock. The image at the right is a petroglyph, an image chiseled directly into the stone. It should be emphasized that neither of the records can be dated; it is not even known which tribes produced them. Archaeological excavations have shown, however, that people did live near the sites of the rock art in the middle of the eleventh century, the time of the explosion in the sky. The photographs appear through the courtesy of the photographer, William Miller.

ask: What did the sky look like in 1054? Was there a bright object near the crescent moon? Was it the supernova?

One of our collaborators works at the US Naval Observatory in Washington, DC, and thus has access to many computer programs involving celestial mechanics. One of them is called "Ancient Moon," and gives a precise determination of the moon's location with respect to the celestial sphere (the background of stars) at any given time on any given date. It turns out that at the time of the supernova the moon wasn't near any of the bright planets. It *was* near the sun, however. Still, the moon is near the sun very often, and I doubt that anybody would have bothered to commemorate an event like that. Finally, as we discovered with pleasure, it was also near the supernova. Figure 2 shows the relevant portion of the sky on July 5, 1054 A.D., the approximate date of maximum brightness for the supernova, as determined by the Chinese records. The coordinates are those typical for astronomy: right ascension (the celestial equivalent of longitude) along the horizontal axis and declination (celestial latitude) along the vertical. The Crab Nebula (or the supernova that produced it) appears at the lower right-hand corner of the field. The position of the moon is shown hour by hour. Note that the moon is a crescent on July 5, 1054 A.D.

Several other circumstances mitigate in favor of the interpretation we have given to the symbols. First, the Chinese and Japanese records of the Crab Nebula supernova suggest that the supernova was about five times brighter than the evening or morning star—the planet Venus in the evening or morning sky. In short, it was extremely bright. Anyone who was accustomed to looking at the sky would certainly have noticed it.

Second, the moon would have been seen in close proximity to the supernova only in western North America. As the diagram shows, this proximity occurs at moonrise. But by the time moonrise had come in China, the moon would have moved four or five degrees in the sky—that is to say, it would have risen in a position different by eight to ten lunar diameters. This is off the diagram, and well away from the supernova. Now the Chinese had a well-developed positional astronomy in the eleventh century; they referred, for example, to a star near the supernova which we call Zeta Tauri. If the supernova and the moon had been very close to each other, the Chinese would have mentioned this in their description. They did not; and in

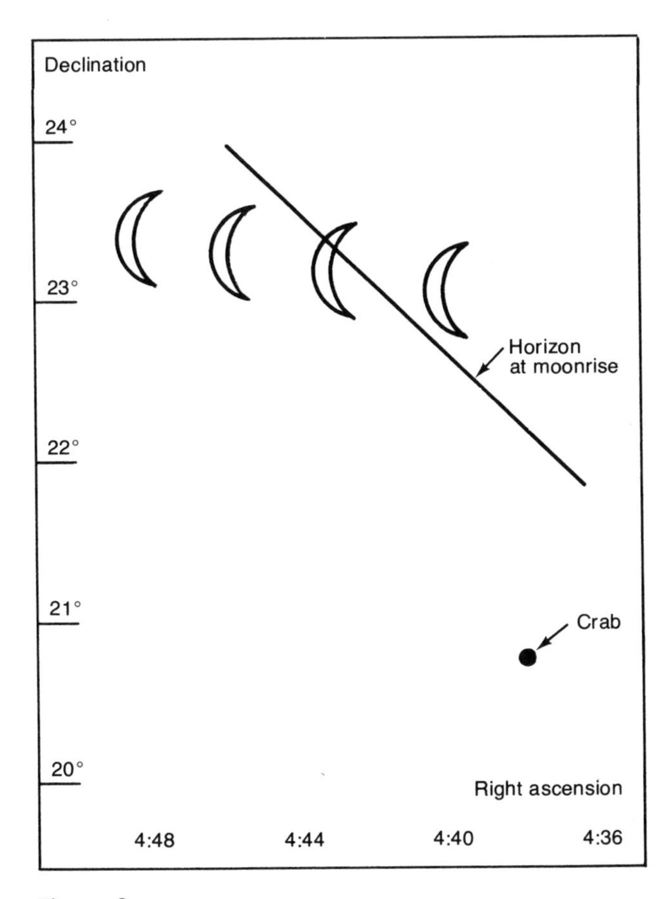

Figure 2
The positions of the moon and the Crab Nebula (or the supernova that produced it), as seen from Fern Cave in northern California on the morning of July 5, 1054 A.D.—close to the day of maximum supernova brightness, according to Chinese records. The coordinate labels for the illustration are those typically used in astronomy; through the application of such coordinates, the stars stand still, and so does the Crab Nebula: it appears at the lower right of the illustration. But both the moon and the horizon move across the unchanging background of the stars. The moon moves from right to left. It is shown at four positions, each an hour later than the preceding one, so that the rightmost crescent shows its position three and three-quarter hours before sunrise, and the leftmost shows it three hours later, and three-quarters of an hour before sunrise. The horizon also moves from right to left, but its sweep across the diagram is far more rapid than that of the moon: it moves five times the horizontal length of the chart in an hour. Accordingly, only a single position of the horizon line is shown: it is frozen at the moment of moonrise in northern California on July 5, 1054. Note the close association of the crescent moon with the supernova at that time.

fact records linking the supernova to the moon only appear, to the best of our knowledge, in western North America, the only place in which the configuration was visible.

Third, every site we have investigated has a view of the eastern horizon, where the events in question occurred. This is not to say that every pictograph or petroglyph that concerns us lies a few feet away from a vista including the lower eastern sky; some of the sites are deep within caves. Still, no site is low in some valley from which the eastern horizon cannot be seen. And even if the site is in a valley it lies on the western side, sometimes on a cliff, so that one can look across the valley and see the sky location at issue.

The cave to which Muriel Kennedy referred in her letter is called Fern Cave. Its entrance is blocked by a wire mesh, for the cave is remote by anyone's definition, and yet some people like to vandalize rock art. One of the pressing reasons for documenting one's discoveries is that sometimes the investigator returns to the site and finds that the art has been mutilated. In some cases, rock art has been used for target practice.

Fern Cave itself is a lava tube that runs for several hundred yards. In one place the roof of the cave has collapsed, producing a small mound of debris on which a colony of ferns about fifteen feet in diameter has grown—thus the cave's name. For almost as far as you can walk, the cave walls are covered with pictographs, including the one mentioned in Mrs. Kennedy's letter. Several of them have been destroyed. Some years ago, for example, a newspaper in the Pacific Northwest decided that the pictographs would be an interesting subject for a series. Photographers sent to the cave found that the pictographs didn't photograph very well. To bring them out, they outlined some of the symbols with chalk.

The records at Fern Cave were probably made with a bit of charcoal—a burnt stick or a burnt bone, perhaps. Figure 3 shows the wall containing the possible supernova record. A close-up photograph (figure 4) shows detail.

Near Fern Cave is a second group of records, at a site known as Symbol Bridge. Although it is exposed to the elements, crescents are still visible on the rock face there, as shown in figure 5. One crescent is situated just above the cross in the photograph; another is at the left of the photograph and is not easily visible; a third is at the bottom. The presence of three crescents

Figure 3
Pictographs at Fern Cave, in Lava Beds National Monument, northern California. Several square feet of the wall of the cave are shown, including a crescent in close proximity to a circle. These are the symbols thought to depict the supernova of 1054 A.D.

is remarkable, for crescents are quite rare: there are probably only thirty or forty "respectable" crescents reported in the entire rock-art literature. Anthropologists who have studied the rock art of California and eastern Nevada, for example, have published a table of 58 common form elements. The list includes circles, dots, wavy lines, and so on. Crescents are not included. New Mexico furnishes a second example. That area is the research province of Colonel James Bain (Retired). He searched for us through his entire collection of 35mm slides, and found only three crescents, two of which we think are relevant to our investigations.

Given these circumstances: that the crescent is a rare element in rock art, that in some records the crescent is placed in contiguity with a bright object, and that three crescents, or very roughly 10 per cent of all the known crescents in American rock art, appear on a single rock face in the same general area, as a crescent-and-bright-object combination—given all this, we are

Figure 4
Close-up of the relevant pictographs of figure 3.

encouraged to believe that the Symbol Bridge site may be another record of the supernova.

We tend also to regard it as a great stroke of luck that the two symbols composing the records which most interest us are different yet easily recognizable, and that one of them is a crescent, rare though they are. One could easily imagine the record as consisting of just two bright-object symbols close together, and in that case, one could spend the rest of one's life trying to figure out the correct interpretation.

The site we consider to be our best lies in northeastern New Mexico in Chaco Canyon National Park. It lies, in particular, at the base of a cliff whose top is occupied by one of the major pueblo dwellings in Chaco Canyon. The dwellings are collectively called Peñasco Blanco—a Spanish name that almost surely has no similarity to the name they had when they were popu-

Figure 5
Pictographs at Symbol Bridge, near Fern Cave in Lava Beds National Monument. The images are weathered, but three crescents can nevertheless be seen. One of them appears at the top of the vertical series of symbols, and has a cross appended to it. The second appears at the bottom of the vertical series. The third, fainter than the others, appears to the left of the vertical series, between the fourth and fifth symbols of that series.

lated. Two pictographs at the base of Peñasco Blanco are of special interest. One is a sun symbol, such as is quite often used to mark sun-watching sites: places where people could line up a pillar or some other nearby fiducial mark with a particular point on the horizon. The other represents the supernova—or we hope it does; and this second symbol appears not to mark a location at all. Instead it is situated in one of the most secure places one can imagine: it appears on the underside of a small overhang. The pictographs in question are shown in figure 6. The small, gourd-like objects in the photograph are nothing archaeological; they are birds' nests. The crescent and supernova make a good approximation of the sky on the morning of July 5, 1054. Notice that the imprint of a left hand is nearby. Among ourselves, we say that it is the artist's signature.

I mention the Peñasco Blanco site in order to point out a big

Figure 6
Pictographs at Chaco Canyon, New Mexico. The camera was pointed upward; the three images at the center of the illustration thus appear in truth on the underside of an overhanging rock. The uppermost of the three is the imprint of a left hand; beneath it are a crescent-shaped symbol plainly reminiscent of the moon and an asterisk-like symbol thought to represent the supernova of 1054. A fourth pictograph, this one a sun-watching symbol, marks a place from which the rising of the sun was lined up on certain days of the year with landmarks on the horizon.

weakness in the kind of work I am describing: We may have a good idea of what the sky looked like when the supernova appeared, and we may establish a date for the supernova in which we can place confidence. What we cannot do with any degree of confidence is date the pictographs and petroglyphs. The carbon in Fern Cave cannot be dated; there isn't enough of it. The records at Peñasco Blanco, which were probably made with hematite, cannot be dated; no method exists. All we can do is determine the time when the nearby civilization was flowering, and argue by inference that this is the most likely time for the record to have been made. To my mind, this is the weakest link in our archaeoastronomical work.

Even so, the available evidence is somewhat encouraging. The chart shown in figure 7 is a tree-ring dating diagram for the Chaco Canyon area. It shows that the years around 1050 A.D. were a time of considerable activity in Peñasco Blanco. This finding is supported by archaeological research in Chaco Canyon; the archaeologists concur that the middle of the eleventh century was its Golden Age. Despite all this, we simply have no way of knowing if 1054 A.D. was the year the putative supernova record was made. Indeed, it is not at all obvious that the supernova should have been recorded as it happened. My own view is that the supernova would most likely have been recorded if it had fortuitously coincided with some other significant event in the life of these people. After all, if you go to Bayeux, France, you can see a tapestry depicting the invasion of England by the Duke of Normandy. Halley's comet also appears on the tapesty, but nobody in his right mind thinks that the people who wove it meant to commemorate that bit of astronomy. It just happened that the comet appeared at a propitious moment and could be interpreted in the context of important political events occurring at the time. My guess is that the likelihood of recording *any* event in the heavens would be enhanced by the occurrence of some remarkable event on earth. Perhaps the supernova of 1054 A.D. occurred simultaneously with a good hunt, a favorable battle, plentiful crops—or a bad hunt, a lost battle, a famine.

Two arguments can be made against our interpretation of the petroglyphs and pictographs I have shown; both are made by anthropologists. One of these arguments is that the records we think commemorate the supernova were in fact only symbols

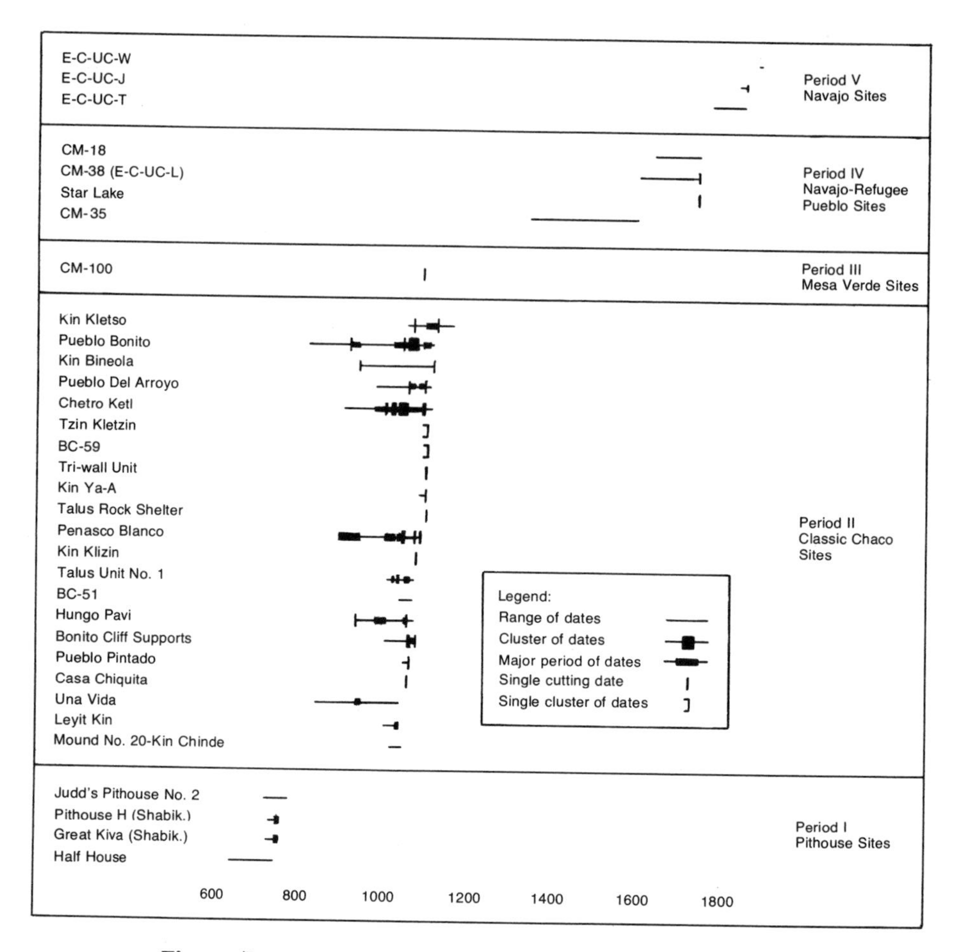

Figure 7
A tree-ring dating of archaeological sites in the Chaco Canyon region of New Mexico. The crux of the technique is simply that the thickness of rings in the trunk of a tree is roughly proportional to rainfall from year to year. Accordingly, core samples taken from the trees in a given region will show a characteristic pattern, and if the rings in some of these cores can be correlated with a particular sequence of years, then older rings within that core—and similar ring-sequences in other trees—can be dated. Proceeding in that fashion, one establishes a dating system that can be used for core samples taken from trees used in the construction of ancient buildings. Peñasco Blanco, a site where crescent-and-bright-object symbols have been found, turns out to lie in the middle of the "Classic Chaco" period. Notice that a "major period" appears at 1054 A.D.

used to mark sun-watching stations. Now we do not dispute the fact that there are sun-watching stations, and that symbols might have been drawn or chiseled to mark them. Still, you'll notice that in Chaco Canyon, for instance, the record that interested us isn't out in the open where anyone could see it. The record is on the underside of a rock face. It is quite hard to find (and I hope it remains hard to find, considering the threat of vandalism). On the other hand, the sun-watching symbol at Chaco Canyon is painted on a vertical rock face and can be seen from anywhere in its immediate vicinity. I should also point out that several of our sites are in caves.

The second objection to our work is an ethnographic one: the argument has been made that the people thought to have lived in western North America at the right time to see the supernova would not have kept records of *anything*. It is certainly true that the Pueblo Indians from whom informants are drawn by anthropologists generally are reluctant to keep records. It is objectionable, however, to extrapolate 900 years from this statement. After all, the American Indian has undergone massive cultural trauma in the last millenium.

The search for supernova records has been fairly productive: we are now aware of over 15 sites spread over western North America, in Arizona, Baja California, California proper, New Mexico, Utah, and Texas. Each such site includes records showing the combination of a crescent and a bright object, and each such record is a reasonable representation of the remarkable events in the sky on the morning of July 5, 1054 A.D. Considering the scarcity of crescents in American Indian rock art, the fact that the close conjunction of moon and supernova was visible only in western North America, and the fact that the archaeological evidence indicates eleventh-century habitation near the sites, we believe we have a strong, if circumstantial, case that the American Indian, when a supernova appeared in the sky, acted much as did the Chinese and Japanese on the other side of the earth: he recorded the event.

For further reading:

"Possible Rock Art Records of the Crab Nebula Supernova in the Western United States," by J. C. Brandt, S. P. Maran, R. Williamson, R. S. Harrington, C. Cochran, M. Kennedy, W. J. Kennedy, and V. D. Chamberlain, in *Archaeostronomy in Pre-Columbian America,* Anthony F. Aveni, editor; University of Texas Press, 1975.

"Rock Art Representation of the 1054 A.D. Supernova: A Progress Report," by J. C. Brandt and R. Williamson, in *Essays in Native American Astronomy,* Anthony F. Aveni, editor; University of Texas Press, 1977.

SHARON GIBBS

THE FIRST SCIENTIFIC INSTRUMENTS

The early history of astronomy has proven to be a field rife with conjecture and speculation; and while it is difficult to find a single reason that will account for this, it seems certain that the nature of the evidence itself has had a great deal to do with it. After all, what remains to us of the science practiced by ancient cultures is just revealing enough to require considerable interpretation, yet this evidence, such as it is, represents our only hope of understanding the origin and extent of early interest in astronomical matters. By reviewing in this article a selection of the material with which the historian of ancient science must work, I hope to communicate some sense of its challenging potential. Unlike other writers in this volume who focus on architectural structures that seemingly were used for astronomical activity, I intend to emphasize small-scale remains. I will consider them in roughly chronological order beginning in prehistory, and I will group them into objects from the Old World (more specifically, the area surrounding the Mediterranean) and the New World (the American continent).

Evidence suggesting a prehistoric interest in astronomy is particularly elusive. In fact, many would contend that no such evidence has been identified. Alexander Marshack disagrees. In *The Roots of Civilization,* published in 1972, Marshack illustrates numbers of curiously marred (or marked) objects found at prehistoric sites in Europe and Africa. Among them is a notched bone tool found on the shores of Lake Edward, one of the headwaters of the Nile River. The tool was found among ruins dating from about 8500 B.C., placing it in the middle of the Mesolithic period. It is notched on three sides (see figure 1), and the notches appear to be arranged in groups. Marshack

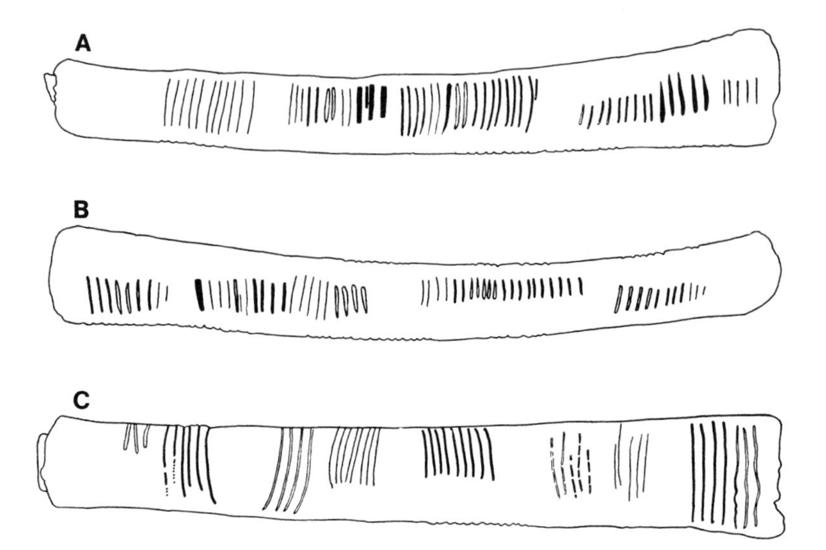

Figure 1
The so-called Ishango bone, an artifact of the Ishango culture, which lived 8,500 years before our era at a fishing village on the shore of Lake Edward, Africa. The bone was evidently the handle for an implement of some sort, possibly a writing instrument, for a small piece of quartz, visible at the left in (A) and (C), is still affixed to the shaft. The bone is most notable, however, for three rows of notches incised into its surface; the notches are further subdivided into groups of varying length and thickness. Alexander Marshack has proposed that the rows shown in (A) and (B) each mark out a total of two lunar months, one scratch added each day. To be sure, an exact lunar month is 29.5 days, but if one looks to the sky and counts the number of days between successive times when the moon is new (that is, invisible), one may count anywhere from 28 to 31 days. Marshack explains the possible significance of the various scratch subgroupings in his book, *The Roots of Civilization,* from which this illustration is adapted.

was unwilling to accept archaeologists' interpretation of these deliberate markings as representing a simple arithmetical game. He considered instead the possibility that the notches were in fact some sort of notation. In particular, he observed changes in both the size and the intensity of the markings and noted that they seemed to have a certain rhythm—or "phrasing," as he put it. He then searched among the phenomena of nature that would have been available to a culture of 8500 B.C., hoping to find the motivation for these marks. He recognized, he thought, similar phrasings in the motion of the moon; that is, he recognized that the intervals of marks on the bone separating the most obvious changes in size and intensity seemed to correspond to intervals of days separating significant lunar events such as new moon, quarter moon, and full moon.

Marshack's identification of the Lake Edward bone as an early scientific instrument well illustrates the role of the investigator's imagination in the interpretation of evidence unaccompanied by written or even pictographic documentation. Remains from a period some 6,000 years after the marking of the Lake Edward bone, and from an area some 6,000 miles closer to the Nile Delta, represent an interpretive problem of an entirely different sort. In 1300 B.C., the decorators of the cenotaph of Seti I, a Pharaoh of Egypt, depicted on the ceiling of the cenotaph an object with an accompanying text discussing "sun," "shadow," and "hour" (figure 2). Twentieth-century excavators recognized a clear relationship between this cryptic but illustrated text and the fragments of objects found in other tombs in the area. The few lines of written evidence and the few fragments of unwritten evidence combined to point to early Egyptian use of the shadow clock: the sundial. But the details of its use were not so readily revealed. Still, information about the duration of Egyptian hours was to be found between the lines—the lines, that is, on the fragments of the putative dials. On the basis of this interpolated data, historians of astronomy were able to conclude that while every Egyptian day was divided into 24 hours, the duration of these hours had no precise definition.

Written and unwritten evidence relating to the early history of astronomy continue to complement each other into the Hellenistic period, when both types of evidence become less fragmentary. An example of the relation between the two can again be found in the practice of marking time by means of changing

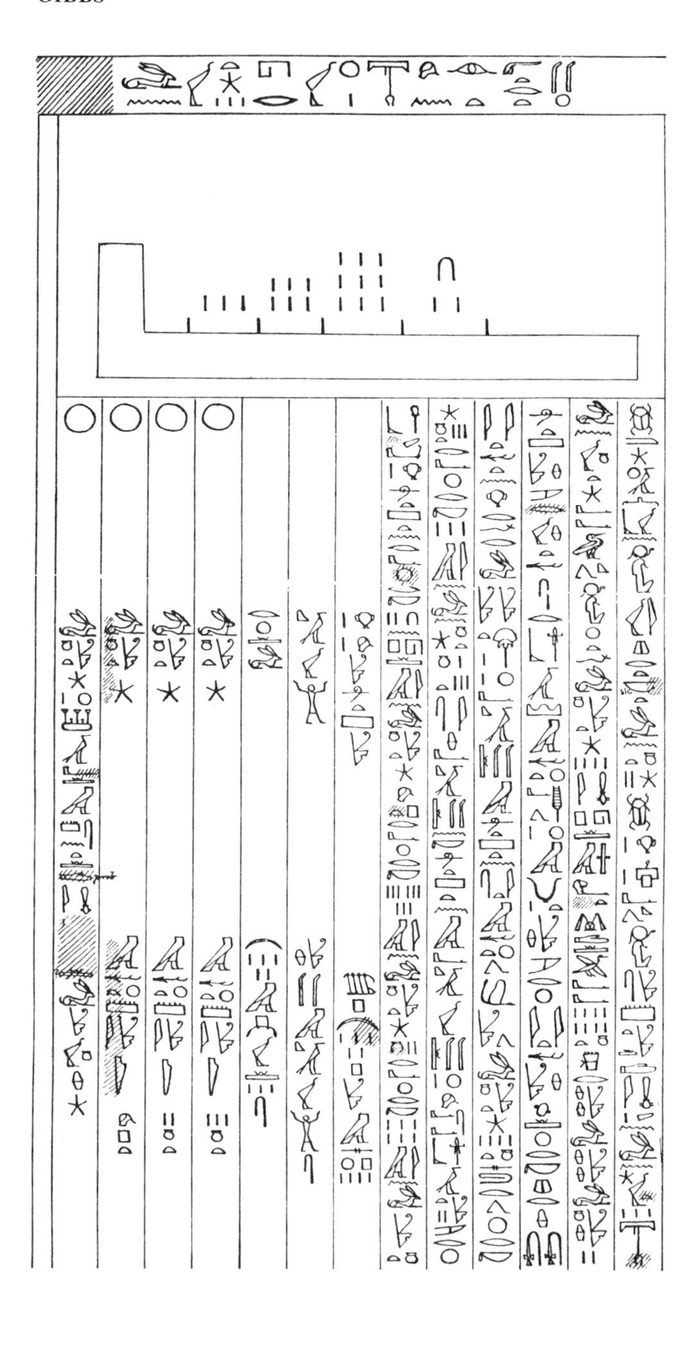

Figure 2
A putative shadow clock, and an accompanying hieroglyphic text, from the wall of the cenotaph of Seti I, a pharaoh of Egypt. The putative clock is the L-shaped object above the columns of hieroglyphics. Above the clock is a single line of glyphs, which reads: "Knowing the hours of the day and night. An example of fixing noon." In the text below the clock, one finds the words "sun" (the combination of a vertical straight line and a circle), "shadow" (the combination of semicircle, vertical line, and fan or sunshade), and "hour" (the combination of rabbit, pot, semicircle, crown, and star); all three appear, for example, in column 10.

shadow lengths, for the architect Vitruvius, writing in about 25 B.C., briefly discussed dial making and listed types of dials that were familiar to him. The names on this list evoke no clear images of the objects to which they refer. Fortunately, however, a number of monumental and portable shadow clocks have survived from antiquity to add meaning to Vitruvius's work. Monumental stone sundials (designed to be used at a single latitude) have been excavated from about 100 Hellenistic sites around the Mediterranean from Spain to the Middle East.

The oldest datable Graeco-Roman sundial (figure 3) serves to illustrate the kind of information conveyed by the artifactual evidence. The dial was found at Heraclea ad Latmum, in what is now western Turkey. As is most often the case with ancient dials, its metal, shadow-casting gnomon is missing. The dial has an inscription on its base that comprises not only a dedication but also an identification of its maker, a third century B.C. craftsman. All of this is unusual: most extant Greek or Roman sundials have no inscription of any sort. Now the precise workmanship evident in the sundial reveals an important Greek refinement of the essential Egyptian technique, for all functional Greek dials show not only the hour of the day but also the day of the year. Both measurements would have been indicated by means of the shadow cast by the gnomon. Thus the Greek sundial is a shadow calendar as well as a shadow clock; its design gives the impression that the Greeks had embodied within it a model of solar motion. In the dial shown in figure 4, concentric lines curving across the sundial's face mark the path of the sun during the day of the winter solstice (upper curve), the spring or fall equinox (middle curve), and the summer solstice (lower curve). The central, vertical line immediately beneath the gno-

Figure 3
The earliest known Hellenistic sundial, from the collection of Greek and Roman Antiquities in the Louvre Museum, Paris. The dial, about one foot in height, is substantially intact, except for its gnomon, the projection point whose sweeping shadow marked the passage of time. The front face of the dial has the shape of a portion of a cone. Seven concentric half-circles are inscribed into its surface. They mark the path of the sun across the sky (that is to say, they mark the path of the shadow of the missing gnomon) on the days when the sun entered each sign of the zodiac in the course of its yearly passage through the sky. The shadow passed along the curve for summer solstice (lowest curve) and the curve for winter solstice (uppermost curve) once each year; it passed along each of the other curves twice. Intersecting all of these curves is a set of lines that divides each day into "hours." (Photograph by Chuzeville, Musée du Louvre)

mon marks the meridian, i.e., noon. The point of the gnomon's shadow in the figure marks the end of the fourth hour on the day of summer solstice.

The relation between the sun's position on the celestial sphere and the lines on a hemispherical dial face is diagrammed in figure 5. Several lines in the diagram connect the points that mark the position of the sun at the end of a certain hour on three days of the year. Such hour-points divide each of the sun's daily paths between sunrise and sunset into 12 equal parts. It is these curves, points, and connecting lines which are projected through a gnomon point onto a smooth surface to create a functioning Greek sundial. The complexity of the resulting projection is dictated by the shape and orientation of the dial face.

Figure 4
The end of the fourth hour on the day of summer solstice is marked by the shadow on a sundial in the Museum of History and Technology in Washington, DC. The dial is a model of an instrument found in Athens, and preserved in the Rijks Museum in Leiden, the Netherlands. The lowest of its day-curves is crossed not only by hour-lines, but also by shorter line-segments that evidently delineated half-hours.

Both written and unwritten evidence combine to establish the fact that in the Hellenistic Age the notion of an hour had become precisely associated with the changing position of the sun. Though the duration of this "hour" varied during the year, being shorter in winter and longer in summer, at least it did so predictably. Both written and unwritten evidence also combine to document the extent of Graeco-Roman knowledge and utilization of projective geometry, for the almost 300 surviving Graeco-Roman dials confirm that the design of an accurate shadow clock and calendar required a fairly sophisticated geometrical sense. It surely comes as no surprise that Vitruvius identifies a number of astronomers and mathematicians as designers of sundials. In Book IX of *De Architectura,* he attributes the so-called quiver design to Apollonius, the mathematician who developed the theory of conic sections. ("Quiver" refers to the appearance of the device: lines flow from its center like

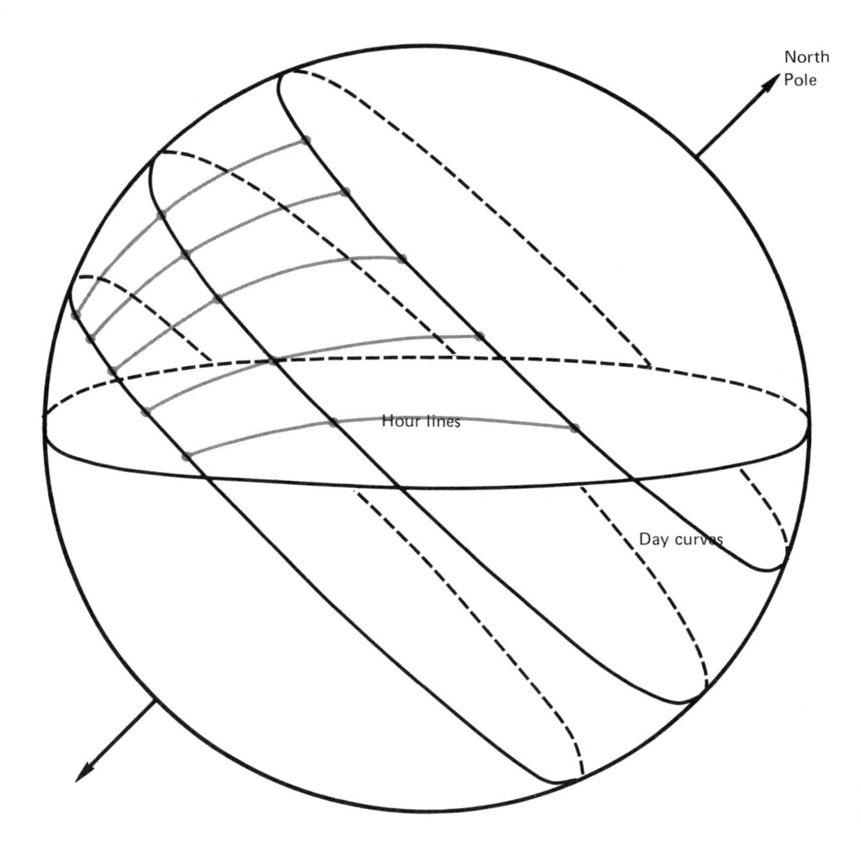

Figure 5
Day-curves and hour-lines on the celestial sphere. The former are the loci that mark the sun's passage through the sky on any one day of the year, such as a day of solstice or equinox. Three such lines are shown in the illustration; they circle the celestial sphere like hoops. The latter are the loci that connect the same hour of the day across several day-curves; five are shown in the illustration. The exact form of the line network projected onto the surface of any given sundial depends, of course, on the shape of the dial. But the general network as shown here, independent of any single projection system, is credited to Eudoxus of Knidos, a fourth-century B.C. astronomer.

arrows spilling from a quiver.) He attributes the planar-disc design to Aristarchus of Samos, who is best known for proposing, long before Copernicus, that the sun was the center of the solar system. Aristarchus's disc was designed to function in the plane of the meridian, with gnomons perpendicular to surfaces facing east and west. Finally, he attributes the day-curve, hour-line network, independent of the specific shape of the dial on which they are projected, to Eudoxus of Knidos, an astronomer of the fourth century B.C. whose cosmological views were adopted by Aristotle.

Vitruvius's list of dials and inventors, together with the quality of the remaining dials, really only hints at the relations that may have existed between the science of gnomonics (or dial making) and other scientific activities in antiquity. The nature of these relations must therefore be inferred. An example of an attempt to do so exists in Otto Neugebauer's proposition (suggested several years ago) that the development of the planar sundial motivated the discovery of the conic sections. His proposed connection has proven difficult to document, partly because very few planar sundials survive. A much stronger case can be made for the contribution of dial making to the sciences of astronomy and geodesy, for the most reliable account we have of the efforts made by Eratosthenes to measure the circumference of the earth reports that his instrument consisted of two hemispherical sundials, one in Syene, the other in Alexandria. On a day when the sun over Syene cast no shadow at noon, the sun over Alexandria cast a shadow that indicated a seven degree difference in inclination. From that observation, and knowing the distance from Syene to Alexandria, Eratosthenes calculated the earth's circumference. We also know from ancient sources that the astronomer Meton set up a sundial on the Pnyx in Athens to aid his fifth-century B.C. researches. Although it has not survived, scholars have suggested that it was similar in design to a well-preserved, very large dial set up just below the Acropolis wall.

The corpus of Greek and Roman sundials contains one example which also seems to document the development of mechanical technology in the last decades of the pre-Christian era. The extent to which it does so depends heavily, however, on an interpretation of some puzzling, unwritten remains within a context of written remains. The unwritten clues can be found inside the Tower of the Winds in Athens. The tower itself, mainly octagonal in shape, is mentioned by Vitruvius, who attributes its design to Andronicus, a Macedonian astronomer. Its outer surface is decorated with fine planar sundials and also with relief sculptures depicting the personified figures of the winds. As for the inside of the tower, nothing remains, except for some curious holes and grooves in the stone floor and walls, to verify its ancient use. Still, the placement of the holes and grooves in the octagonal part of the structure, and also in a semicylindrical annex to the main octagon, convinced Derek Price and Joseph Noble that the tower once housed an elaborate water clock of a type known to Ktesibios, Philon, Heron, and Vitruvius. Price and Noble propose that the upper and lower levels of the semicylindrical annex housed a tank which provided a constant head of water, and beneath it a cylinder containing a float. Water in the upper tank fed in a slow drip into the lower cylinder, causing the position of the float to vary at a constant rate. A chain attached to the float ran by a series of pulleys to a weight, which in turn governed the motion of the clock mechanism proper. Figure 6 shows their reconstruction of the clock, no part of which survives.

The visual device used by Andronicus to display the ancient, seasonally variable hours in the Tower of the Winds is completely open to conjecture. Price and Noble tend to view it as consisting of an anaphoric clock face like the face described by Vitruvius and preserved, in principle at least, in examples of the planispheric astrolabe (figure 7). An anaphoric clock face consists of a solid, flat disk containing a stereographically projected star map. A fragment of just such a large, flat disk engraved with a star map was found in Salzburg in 1902. It lent convincing evidence to the reconstruction, which includes the ecliptic or zodiac circle—the path taken by the sun through the background of stars in the course of a year—fitted with a moveable marker representing the sun, all of this viewed through a network of stereographically projected hour-lines and day-curves, just like those invented by Eudoxus for projection onto the face of a sundial. When the clock was operating, the flowing water, the pulleys, and the weights combined to cause the star map, with its sun symbol, to rotate once every 24 hours behind the fixed hour-line network.

To be sure, the restorers of the clock in the Tower of the Winds, in comparing its visual device to a planispheric astrolabe,

Figure 6
The water clock in the Tower of the Winds, Athens, as reconstructed by Derek J. de Solla Price and Joseph Noble. According to the two, the semicylindrical structure at the left contained a water tank and a series of floats, pulleys, and weights, which generated a force to drive a clock in the octagonal structure at the right. (Illustration © National Geographic Society)

have bridged a considerable amount of time, since the earliest surviving instrument of the type pictured in figure 7 dates from the tenth century A.D., almost 1,000 years after Andronicus completed his work. The foundation for this 1,000-year bridge lies in the writings of Claudius Ptolemy, an astronomer of the second century A.D. who made crucial contributions to the advance of astronomy, yet himself acknowledged a considerable debt to his predecessors. In a treatise on astrological principles, Ptolemy refers to an object he calls an "horoscopic instrument," and says that it contains a "spider" which marks the fixed stars. In another essay, "On the Planispherium," he describes the method and utility of stereographic projection. These two pieces

Figure 7
An astrolabe made in western Europe in 1542, and now in the collection of the National Museum of History and Technology, Washington, DC. Price and Noble believe that the clock face in the Tower of the Winds had a similar nature. An astrolable in general comprises two metal plates, one atop the other. The outer of these is a star-map in the form of a skeleton network; the dagger-like points extending from the skeleton-like scaffolding in the illustration show the positions of prominent stars such as Aldebaran, Regulus, Arcturus, Spica, and Vega. Beneath the network is a solid plate with hour-lines and day-curves inscribed in its surface. Other lines engraved on the upper half of the solid plate are the stereographically projected elements of an earth-centered co-ordinate system including attitude circles parallel to the horizon and azimuth circles which pass through the zenith.

of evidence have suggested to scholars that Ptolemy was familiar with the astrolabe, in the sense of a skeleton star map which could be rotated above a solid plate engraved with day-curves and hour-lines for a specific geographic latitude. Obviously a certain amount of inference is necessary if one is to place a two-part "horoscopic instrument" in Ptolemy's hands. The evidence only indicates that he could have constructed and used an astrolabe as we know it.

The thousand-year gap between Ptolemy's reference to an horoscopic instrument and the earliest surviving astrolabe has been explained, incidentally, by the same factor which explains the rarity of metallic instruments in general: i.e., lack of materials. At times when new metal was generally unavailable, the old pieces (including most of the bronze gnomons of monumental sundials, all but a handful of the small portable dials, and apparently all of the "horoscopic instruments") were pressed into service.

Whatever the scarcity of the evidence, the task of interpreting it in historical times is made easier when the written and unwritten evidence complement one another, as has been the case in the examples discussed so far. One cannot always expect to find complementary evidence, however; and indeed, isolated bits of unwritten evidence as puzzling as those from prehistoric times also turn up in later periods. One such bit of evidence went without a complete interpretation for almost 75 years before its relatively sophisticated mechanical characteristics were recognized. In a way, this is easily understandable, for the device seems quite extraordinary when it is placed among sundials, water clocks, and horoscopic instruments.

In the spring of the year 1900, a group of sponge divers was forced off its normal course by bad weather. They ultimately dropped anchor off the island of Antikythera, situated in the Mediterranean between Crete and the mainland of Greece. The divers had never examined this area for its sponge fishing possibilities, so they made the best of their situation and did so. In the course of this exploration, they found the ancient wreck of a large ship. Within it were several bronze and marble objects, which they brought to the surface. And among those objects were four bronze fragments, apparently related to one another, but all heavily corroded. They were eventually dated to 87 B.C. During the years that have followed their discovery, scholars have determined that these fragments were part of a geared mechanism built on a series of bronze plates. Gamma-ray photographs taken in 1971 confirmed this determination.

The photos also led to an interpretation of the significance of the gearing. In 1974, Derek Price reported that "The [Antikythera] mechanism can now be identified as a calendrical Sun and Moon computing mechanism, perhaps made by a mechanician associated with the school of Posidonios on the Island of Rhodes . . . The most spectacular aspect of the mechanism [Price continued] is that it incorporates the very sophisticated device of a differential gear assembly . . . [Before this interpretation, the differential gear was thought to have originated just previous to the Renaissance.]" In a reconstruction of the gear system, shown in figure 8, the differential turntable is visible as the largest gear element on the back of the instrument. According to Price's interpretation, a single turn of the drive wheel simultaneously introduced two separate rates of rotation to the differential turntable. Thus 19 rotations of the drive wheel introduced both 19 and 254 rotations to the differential turntable, which, as a result, moved through (254-19)/2 revolutions. The number of revolutions introduced to and resulting from the differential turntable are relevant to the Metonic cycle, an early recognition of the fact that a given phase of the moon (say full moon) falls on the same calendar date (say December 12) only once every 19 solar years (or 235 lunar months). Price proposes that the Antikythera mechanism is designed to mimic the Metonic cycle as a physical theory and not to simply calculate from it or produce the right appearance.

The unwritten evidence I have discussed up to this point, beginning with the Lake Edward bone and ending with the Antikythera mechanism, suggests the kinds of instruments that were designed or used by scientists in the Old World, from prehistoric times to the Middle Ages. Plainly it points to a sophisticated understanding of periodic astronomical phenomena, and plainly it indicates that by the beginning of our present era, residents of the Old World were adept at applying the mathematics of gnomonic projection, stereographic projection, and gear ratios. We have every reason to suspect that the instruments which survive played a role in the development of Old World science. Certainly the careful interpretation of these in-

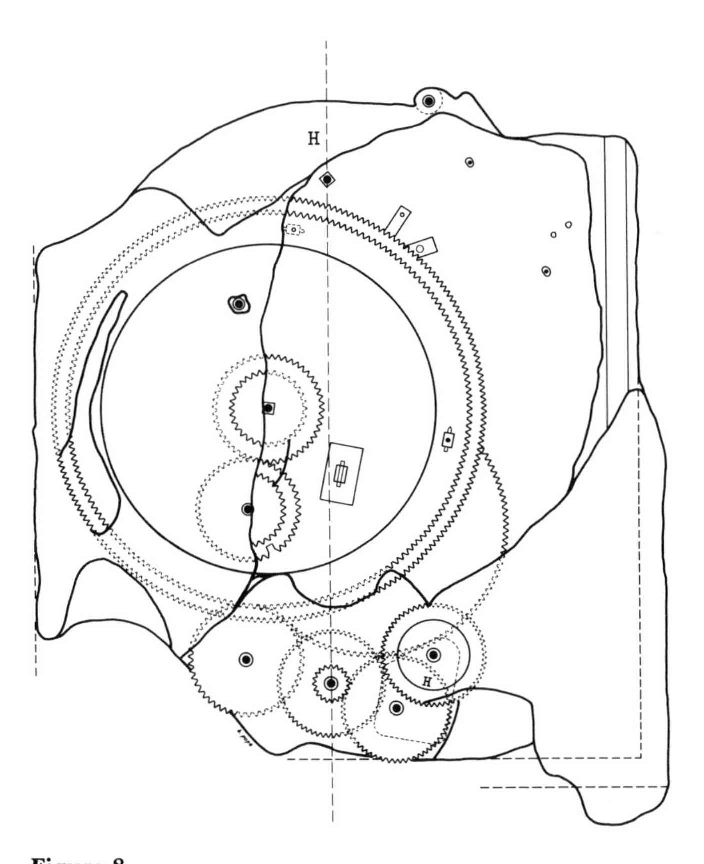

Figure 8
The gear assembly of the Antikythera mechanism, as reconstructed by Derek J. de Solla Price. The gears themselves, as discovered by sponge divers in 1900, were simply a set of heavily corroded metal objects. But Price, basing his conclusions in part on the numbers of teeth in the various wheels, determined that certain aspects of their revolutions correspond to certain aspects of the Metonic cycle, an attempt to reconcile the passage of solar and lunar time. The number 235, for example, is embodied by the gears, and the Metonic cycle is 235 lunar months long; that is to say, 235 lunar months is a lowest common denominator in the periodicities of the sun and the moon.

struments can reveal much about the scientific knowledge of their makers.

The wealth of information supplied to us from Old World sources seems all the more valuable when it is compared to what little is known about the history of astronomy in the New World. Consider the Tower of the Winds. Even if we have essentially no direct evidence to suggest that it once contained a scientific instrument, we have at least the sundials on the outside of the structure. Accordingly, no flight of imagination is required to suppose that the tower served an astronomical purpose. Compare this with the kinds of clues offered by seemingly analogous structures known from Mesoamerica. Figure 9 shows a tower in Palenque, Chiapas, Mexico, that modern-day Palenqueños are

Figure 9
The tower at Palenque, Chiapas, Mexico, believed to have been used in the seventh century A.D. by Maya astronomers. This belief, however, is based entirely on a hieroglyph, possibly astronomical, that decorates its inner walls. The site in general appears to have been occupied from the sixth through ninth centuries; we have little to go on in identifying it as a palace, but scholars usually do so, if only because its rooms appear to have been designed to be lived in, and lack much of the ceremonial aspect of other buildings—most notably, the Maya pyramids. Inscriptions found at the site have made it possible to identify by name the ruling dynasties associated with most of the architectural remains.

convinced was used by astronomers in the eighth century A.D. What clues does it contain to suggest its former use? A hieroglyph associated with the planet Venus decorates its inner walls, but aside from that there is nothing. While the Venus glyph standing alone on the wall of the Palenque tower clearly indicates a Maya fascination with the planet Venus, it cannot convey involvement in astronomical activity as convincingly as does, say, a functioning sundial. The designer of a functioning sundial had to be engaged in astronomical activity. Archeoastronomers can only speculate on possible Maya astronomical activity suggested by the tower's placement and orientation.

A second Mesoamerican example, a round tower at Chichen Itza, Yucatan, Mexico, was identified as an astronomical observatory in 1875, but speculations on the appropriateness of this designation were not carefully analyzed until a century later. What clues now remain? Nothing more than three asymmetrically oriented windows at the top of the tower, and a set of asymmetries in the overall plan of the building—primarily in the orientation of the bases on which the tower was built. Figure 10 shows the earliest known photographs of the windows at the top of the tower. They aren't perpendicular to one another, or at 45-degree offsets. Indeed, they exhibit no obvious mutual relationship at all. The bases, too, face various directions without an obvious geometrical significance. However, the study of all the asymmetries has revealed that several of them are directed toward important setting positions of Venus, including the maximum northerly and southerly such points. These directions appear not only in the windows but also in the base. In particular, the direction associated with a lower stairway in the structure points to the maximum northerly setting position of Venus. Now there are no hieroglyphs depicting Venus on this tower, but the apparent Venus alignments take on added significance when they are considered in the context of Maya religious thought and cosmology. We know from ethnographic and documentary evidence that the Maya were interested in the planet Venus; and we know that they were particulary concerned about its appearance on the eastern horizon after it had disappeared from the sky in the west. We also know that round towers were associated with the Venus god in his capacity as god of the wind. Such are the clues that historians of astronomy must work with in the New World.

Figure 10
The earliest known photographs of the windows at the top of the Caracol, Chichen Itza, Yucatan, Mexico. The images, taken by Carnegie Institution archaeologists in 1920 (before they began a restoration of the site), show a set of windows at a curious mutual orientation, suggesting no purely aesthetic and no geometrical significance. However, several of the orientations of windows and other parts of the structure seem linked to positions at which the planet Venus sets on the horizon.

Graphic or textual evidence is extremely important in New World archaeoastronomy, for without it, no conclusions can be drawn concerning the use of instruments by New World astronomers. The reason is that no physical objects remain in the New World, awaiting interpretation in astronomical contexts—no physical objects except the ruins of buildings. To be sure, the graphic evidence is relatively rare and easily misunderstood; perhaps the most well-known example appears in the Codex Bodley, painted in Oaxaca in about 1520, on the eve of the Spanish conquest. The Codex apparently depicts the dynastic history of the royal family in the district from the time of its mythological origin in 692. Among the pictorial conventions is a temple studded with small circles that look like light bulbs (figure 11). The circles have been identified as the convention for stars. Within the temple is the figure of a Mixtec Indian viewing the horizon over a pair of crossed sticks. The crossed sticks turn up elsewhere in the Codex, sometimes framing the

Figure 11
A figure of possible astronomical significance taken from the Codex Bodley. The codex was painted by the Mixtec Indians of central Mexico in about 1520 A.D., just before the Spanish conquest. Like other codices, it is a book of sorts, or more precisely a set of pages linked like the folds of an accordion. A typical codex would have ten or more such pages, with each page containing four tiers of perhaps ten pictographs. In sum, then, the codex would contain several hundred images arranged in sequence. Mixtec codices, unlike those of Maya origin, are entirely pictographic; they contain none of the more symbolic hieroglyphs, except for the occasional identification of a human figure by a symbol otherwise used to represent a day in the Mixtec calendar. The Indian shown in the illustration is part of an image perhaps three-quarters of an inch tall. He is given no name, but he appears to be a priest, looking out from the window of a temple. The light-bulb-like figures studding the temple are thought to represent stars; and the priest appears to be using crossed sticks as a collimating device to aid him in his observation of something at the horizon.

symbol for star. Accordingly, it is tempting to recognize in this simple device just the kind of instrument needed to aid observations of astronomical phenomena occurring near the horizon. But it is also true that astronomy is not the only context in which the crossed sticks appear. Indeed, the sticks sometimes seem to serve merely as decorative devices.

The problem of interpreting a symbol such as crossed sticks is shared by a number of even less certain cases. Among these is the claim that a Mesoamerican shadow clock is represented in a hieroglyphic figure which decorates the temple of Venus at Xochicalco in central Mexico (figure 12). The glyph itself con-

Figure 12
An image claimed to be a Mesoamerican shadow-clock, from Xochicalco in central Mexico. The image in question is engraved on the west face of the Temple of the Plumed Serpent, south of its central stairway. The top part of the image has been described as "one trapeze crossing another." It has been claimed, moreover, that one of the trapezes was oriented north to south, the other east to west. If so, the shadow of the north-south trapeze would move from west to east in the course of each day. The site of Xochicalco flourished in the period between 800 A.D. and 1520 A.D. under the influence of Toltec, Zapotec, Mixtec, and Maya peoples.

sists of a quartered disk surmounted by a sort of crown; the crown in turn consists of two interlocking arches perpendicular to each other. As for the interpretation of all this, the disk has elsewhere been identified as a figure of the sun; the "crown" in other contexts seems to refer to a calendar year. The identification of a sundial is made despite the lack of convincing evidence that Mesoamericans would have been interested in the information supplied by such an instrument.

I doubt that further examples are necessary to establish the difference between Old and New World evidence regarding the use of instruments in ancient astronomy. Clearly each site offers distinct challenges, yet more than a few scholars in the field have decided that the interesting challenges exist in the Old World and not in the New. It is true that the Mesoamerican material does not yet seem to contain a critical mass of clues. But as investigators turn in greater number to the study of New World astronomy, there is an increasing chance that mutually supportive conclusions will converge to create a coherent picture of astronomical activity by the ancient Americans.

For further reading:

The Roots of Civilization, by Alexander Marshack, McGraw Hill, New York, 1972.

The Cenotaph of Seti I, by H. Frankfort, Memoir 39 of the Egypt Exploration Society, London, 1933.

Greek and Roman Sundials, by Sharon Gibbs, Yale University Press, New Haven, 1976.

"The Astronomical Origin of the Theory of Conic Sections," by Otto Neugebauer, *Proceedings of the American Philosophical Society,* **92,** 136–38 (1948).

"The Water Clock in the Tower of the Winds," by J. V. Noble and D. J. Price, *American Journal of Archeology,* **72,** 345–55 (1968).

Gears from the Greeks, by Derek J. de Solla Price, Science History Publications, 1975.

"The Caracol Tower of Chichen Itza—an Ancient Astronomical Observatory?", by A. F. Aveni, S. L. Gibbs, and H. Hartung, *Science,* **188,** 977–85 (1975).

"Astronomical Signs in the Codices Bodley and Seldon," by H. Hartung, in *Native American Astronomy,* A. F. Aveni, editor, University of Texas Press, 1977.

"Crossed trapezes: A pre-Columbian astronomical instrument," by A. Digby, in *Meso-American Archaeology: New Approaches,* N. Hammond, editor, University of Texas Press, 1974.

ANTHONY F. AVENI

OLD AND NEW WORLD NAKED-EYE ASTRONOMY

" . . . in studying the convoluted orbits of the stars, my feet do not touch the earth, and seated at the table of Zeus himself, I am nurtured with celestial ambrosia."

Though Ptolemy wrote that phrase two millenia ago, it no doubt aptly expresses the feeling any ancient sky watcher would have experienced when he turned his gaze to the stars long enough to see the sublime precision of celestial motion unfold. For modern folk the majesty of the firmament is unveiled only through the mastery of a complex instrumentation; the practice of astronomy has long since ceased to be a matter of naked-eye observation. Dependent upon our machines, yet awed by the remains of the ancient world, we ask, How could our forebears have constructed the pyramids, erected the statues on Easter Island, or carved the Olmec heads without technological assistance? How could they have made their scientific discoveries without instrumentation? Some of us feel compelled to attribute their success to outsiders, ancient astronauts who long ago traversed the galaxy bearing us the gift of a great science and technology which has since vanished from the pages of history.

But the loss may be in ourselves. Ancient records tell us that our predecessors scaled great astronomical heights. Because the heavens were a part of their lives, they labored attentively to follow their gods and goddesses, who were symbolized by the sun and moon, the planets and stars. They enjoyed an intimate contact with nature—a contact which our technology makes difficult for us by fostering the artificial environment in which we play out our lives. Indeed, ancient astronomers were nurtured with celestial ambrosia only because they pulled up to the table and helped themselves directly. In the Old World of the Mediterranean, they created most of the astronomy with which we as historians are familiar. On the American continent, other races, entirely separate from those of the Old World, also created a sophisticated system, an astronomy of equal brilliance.

It is instructive to contrast certain aspects of the Old and New World systems in order to show how naked-eye observations of the heavens, at times interpreted similarly, at other times differently, produced models of the universe for the inhabitants of both sides of the Atlantic. These models, often laced with complex geometry and numerology, were used to predict the course of certain astronomical events with surprising accuracy. Because space does not allow me to review the detailed astronomical progress of both cultures, I shall utilize selected examples, such as the study of lunar eclipses and the cyclic motion of Venus, to demonstrate the quantity of detailed information that can be extracted from simple observations. I shall also try to show how and why the mental frameworks engendered by these astronomies influenced the way people viewed the cosmos.

Why did all ancient civilizations look skyward early in their development? Surely chronology made the earliest demands upon astronomers of antiquity. Ancient hunter-gatherers kept close watch on the changing appearance of the sun and moon. For them, these luminous disks were created expressly for the keeping of time. They were the primal source of light, of the tides, of body rhythms, of the seasons. In particular, two basic cycles, one of them comprising the phases of the moon, represented by successive first appearances of the thin crescent in the west after sunset, and the other the annual solar cycle, most easily discerned by following the daily progress of the sun on the horizon at dawn and at dusk, were found to be represented approximately by $29\frac{1}{2}$ and $365\frac{1}{4}$ day periods.

Our minds constantly seek to make order out of chaos, so it is not surprising that the first astronomers felt compelled to fit the solar and lunar rhythms together. But an integral number of lunar-phase cycles will not divide exactly into the tropical year. A round of 12 moons falls 11 days short and 13 moons is 19 days too long.* Accordingly, we find many early chronologers alternating between years of 12 and 13 months. After a

*This seems not to have bothered the Romans, who handed down our modern calendar of 12 fixed months (a contraction of moon-ths) per year—a time scheme which bears no sensible relation to the events it purports to represent. Already the civilized world was beginning to detach itself from the natural environment.

period of 19 years—the so-called cycle of Meton, devised by that Athenian astronomer in the fourth century B.C.—the two were found to fit together more perfectly. In a sense, Meton had discovered a large gear in the universal timepiece which meshed perfectly with two smaller gears. The attempt to group together shorter periods to make longer ones seems to have been a common characteristic of early astronomers.

But why did the astronomers strive for precision? Perfection was unnecessary in meteorological or agricultural prediction; the answer lies in the spiritual realm. Religious ceremonies always marked the completion of any natural cycle, and early astronomer-priests believed the deities who represented the forces of nature, both physical and social, demanded a strict observance of the rites. When all was in order in the heavens there would be peace on earth, but a disorder necessitated careful corrections in the calendar, lest the timing of good and bad omens occupy the wrong place in human history. Consequently, better observations were in great demand. The priests attended meticulously to the needs of the calendar. As far as the people were concerned, the source of power lay in the priests' hands.

In Bronze-Age Britain, monumental megalithic architecture was created to chart both primary celestial deities: the moon and the sun. And though they paid little attention to lunar excursions, New World astronomers were encapsulating the movement of the sun in their earhworks with remarkable accuracy (figure 1).

The Hopi-Navajo of Arizona denoted important days in the solar year by fixing the position of sunrise and sunset on prominent peaks and notches in their landscape. The ancient Maya of Yucatan, probably the cleverest astronomers in the Americas, incorporated the sun's movement into their architecture by way of an ingenious "solar observatory" which they erected about the beginning of the Christian era. Standing atop pyramid E-VII at Uaxactun, the astronomer-priest could see the sun rise on different days of the year over a cluster of three temples located opposite an open plaza to the east of his vantage point. Embedded in dense rain forest, these buildings were carefully erected on artificially elevated mounds. The top of the northernmost structure precisely marked the position of sunrise on the longest day of the year, June 21, when the sun reaches its most northerly sunrise position. Conversely, the southernmost

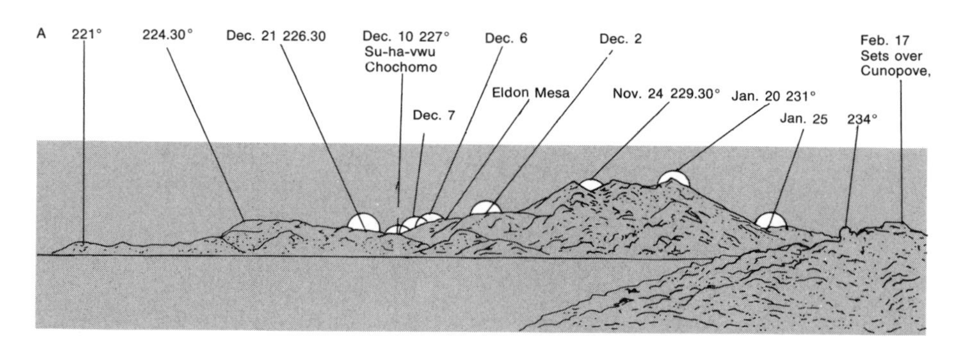
A
221°
224.30°
Dec. 21 226.30
Dec. 10 227°
Su-ha-vwu
Chochomo
Dec. 6
Dec. 2
Feb. 17
Sets over
Cunopove,
Eldon Mesa
Nov. 24 229.30°
Jan. 20 231°
Dec. 7
Jan. 25
234°

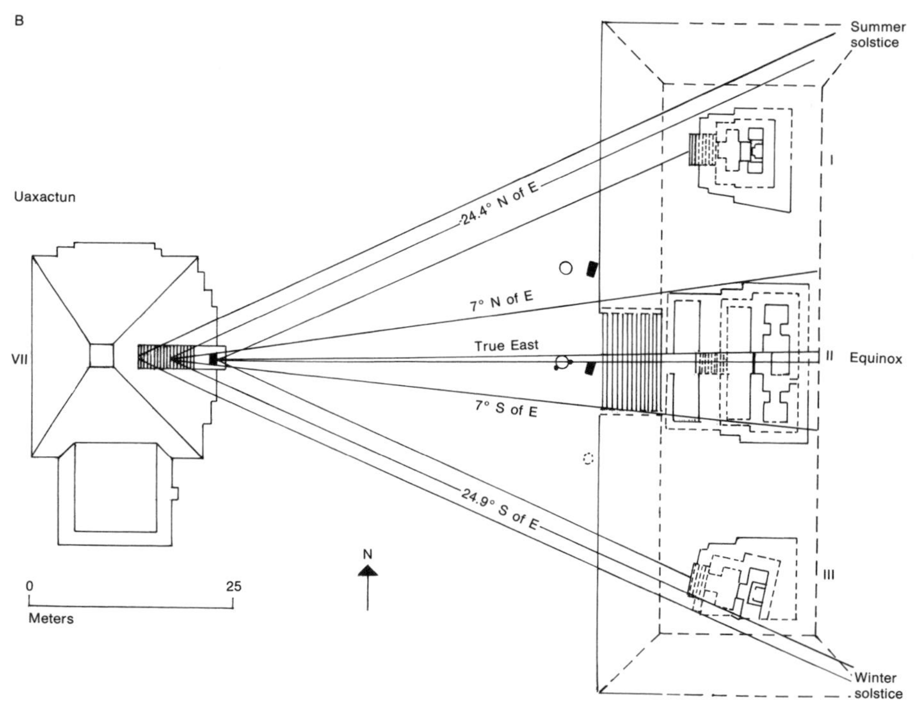
B
Summer
solstice
Uaxactun
24.4° N of E
I
7° N of E
True East
VII
II
Equinox
7° S of E
24.9° S of E
N
0
25
Meters
III
Winter
solstice

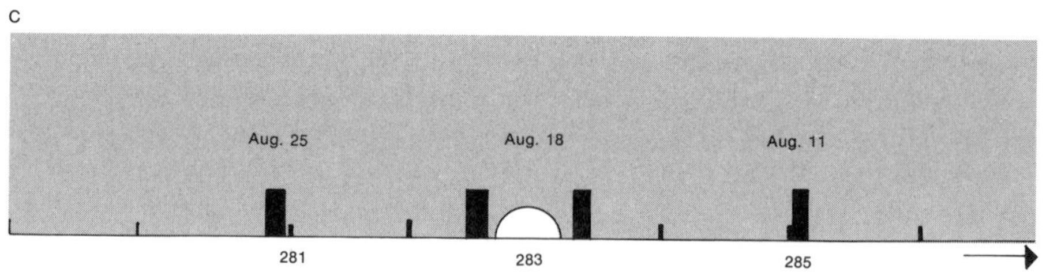
C
Aug. 25
Aug. 18
Aug. 11
281
283
285

building fixed the sunrise on the shortest day of the year, December 21. These standstill positions, called solstices by the Greeks, who also watched them, must have been important time markers for the ancient Maya. The sun was watched with great care as well on the equinoxes (March 21 and September 20), when the central building precisely marked its rising position. On these two dates, the sun rises exactly in the east and sets due west. Daylight and darkness are of exactly equal duration.

What devices were used to establish these sun alignments? We think ancient American astronomers employed crossed sticks to register the position of the sun precisely, so that the architecture could be aligned accordingly. Codices (that is to say, picture documents) surviving from before the time of the Spanish conquest reveal astronomers posing in the doorways of buildings, their instruments in hand, aligning their gaze toward distant objects along the horizon (figures 2 and 3). A crossed stick mounted atop a pole (or perhaps a natural marker in the landscape) would serve as a foresight; and an ancient astronomer could then plant a stick as a backsight and fix a solar alignment by observing the sunset on a given day precisely through both notches. Permanent markers could later be used to fix the positions of the two locations, and on successive days, the astronomer would adjust one of the sticks and mark new alignments to follow the changing positions of the sun at the horizon throughout the year. The return of the sun from the same direction to its original position would indicate that a solar-year cycle had been concluded. The same principle would apply to

Figure 1
Three New World sun-watching schemes, all of them involving the sun at the horizon. A calendar in the landscape is shown in (A). The Hopi-Navajo marked important days in the year by memorizing the positions of prominent notches or peaks on the local horizon where the sun stood on important dates. Part (B) depicts Uaxactun Group E, a Maya solar observatory. An observer situated on the large pyramid marked the sunrise over the tops of three eastern temples precisely placed to mark the solstices and the equinoxes. A way in which the horizon might be used as an agricultural timetable is suggested in (C). The Incas of ancient Peru built cylindrical towers along the horizon of Cuzco, their capital. When the setting sun arrived at a particular tower, the planting time for a certain elevation was said to be correct. Arrival of the sun at another tower signified that crops should be sown at a different elevation.

Figure 2
The use of crossed sticks to fix the positions of astronomical objects at the horizon. This drawing shows an astronomer with crossed sticks perched in the doorway of a temple. Stars studding the outside of the structure give it a special astronomical significance. The image is taken from a Mesoamerican picture document that survived the Spanish conquest. (For a photograph, see figure 11 of Sharon Gibbs' chapter in this volume.) Mesoamerican architects often oriented temples preferentially so that they would align with specific astronomical events at the horizon.

the moon and the planets, which describe more complicated horizon periodicities.

How accurately can one determine celestial position by such a technique? Suppose the foresight and backsight are separated by a distance of 500 meters. If an angular error of one-quarter degree (about half the angular diameter of the sun) is allowed, then the backsight can be laterally misaligned by as much as two meters without transcending the margin of error, or, for a pair of sighting sticks at one kilometer separation requiring an accuracy of one solar diameter, the lateral displacement is eight meters. Since the daily shift of the sunrise and sunset positions along the horizon is more than one-quarter degree (except near the solstices), we see that the simplest naked-eye observations in the natural or architectural environment could easily be used to give a day-to-day tally of the motion of the sun throughout the year.

In Cuzco, Peru, the Inca monitored the sun's movement during the most crucial times of the year by erecting a set of four

Figure 3
A second example of the Mesoamerican crossed-stick symbol. Here an eye-and-stick combination forms part of the headdress of an astronomer, who converses with another wise man about the state of the universe.

cyclindrical towers, or *mojones,* on the highest hill to the west of their capital (this is shown in figure 1C). The astronomer performed his duty atop the *ushnu,* a pyramid located in the middle of this ceremonial center. For there he viewed the sunset on its daily southward progress during August, thus forecasting the arrival of spring in the Southern Hemisphere. Spanish chroniclers living in Cuzco shortly after the Hispanic invasion tell us that when the sun touched the northernmost tower, it was time to begin to plant maize in the highest elevations surrounding the valley of Cuzco.

Days later, when the sun's image was framed by the pair of towers in the middle of the array, farmers within the valley of Cuzco started their planting. When the sun returned from the south, the sun towers served a similar function, this time to set the intervals within which to harvest the crops. The Inca thus saw in their landscape an accurate agricultural timetable. As best we can judge from old maps and documents the Spanish produced about Cuzco, the key date must have occurred about August 18. Unfortunately, nothing remains of the towers today. The Spanish conquerors dismantled them stone by stone in the sixteenth century for use in the construction of aqueducts in their new city.

The Indian civilizations of the Americas flourished in the tropics. Here the sky takes on a different aspect from that in the more temperate climates where the Classical civilizations developed. Therefore, we might anticipate that the astronomies of these diverse cultures would exhibit some differences. For example, Inca, Maya, and Mexican astronomers noted the dates of passage of the sun across the zenith, a phenomenon which never occurs in the temperate zones, and in much of Central America these dates served a meteorological function, foretelling the start of the rainy season. One seventeenth-century Yucatecan writer, Juan Pio Perez, attaches a further significance to the zenith days in the calendar which his native ancestors had used 400 years earlier:

"Our progenitors [he reports] sought to make the New Year begin from the precise day when the sun returns to the zenith of this peninsula on his way to the southern regions, but being destitute of instruments for their astronomical observations and guided only by the naked eye, erred only 48 hours in advance. That small difference proves that they endeavored to determine with the utmost attainable accuracy the day on which the luminary passed the most culminating point of our sphere and were not ignorant of the use of the gnomon in the most tempestuous days of the rainy season."

In ancient Greece, astronomers marked the progress of the sun by following the shadow of a vertical stick or gnomon on a calibrated surface. (See the chapter by Sharon Gibbs in this volume.) But we cannot say for sure whether shadow-casting techniques with a gnomon were important in America. Evidence from New World archaeoastronomical studies suggests that it is more likely that the image of the sun was received through a vertical hole incorporated into a special sun temple. Such a device permitted a shaft of solar light to pass into a darkened observation chamber built into the structure. Consider Structure P at Monte Alban, in modern-day Oaxaca, about a day's drive from Mexico City (figure 4). Though the structure was built more than 2,000 years ago, the image of the sun still passes through such a tube onto a small altar on the two days of the year, May 8 and August 5, when the sun casts no shadow at noon. This zenithal sight tube is fashioned so as to receive an image of the sky $1\frac{1}{2}°$ wide. Thus it will admit at least a portion of the sun a few days before or after a true zenith passage. However, by experiment we find that the sun's disk can be

Figure 4
Observing the sun when it is high in the sky. Tropical astronomers attached great importance to the days when the sun passed directly overhead at noon. In the structure shown here (Structure P at Monte Alban), they noted this time by receiving the light of the sun through a vertical tube incorporated into the monument.

precisely centered in the field of view only on the days of zenith passage.

Across the plaza, another astronomer may have stood in the predawn twilight on a day of zenith passage, crossed sticks in hand, in a small room atop Structure J, which was deliberately skewed from all other buildings at the Zapotec hilltop site. The line of his gaze, directed by the orientation of the door jamb, would pass directly over the access chamber to the solar observatory, reaching the eastern horizon precisely at the place where the bright star Capella rose to announce the crossing of the sun in the zenith; for, at the latitude and building time of Monte Alban, Capella made its first annual appearance in the predawn eastern sky (its heliacal rising) on the same day the sun underwent its first zenith passage.

The alignments at Monte Alban and Uaxactun reveal that the ancient Mesoamericans had a penchant for utilizing architecture and the landscape to follow the progress of celestial luminaries;

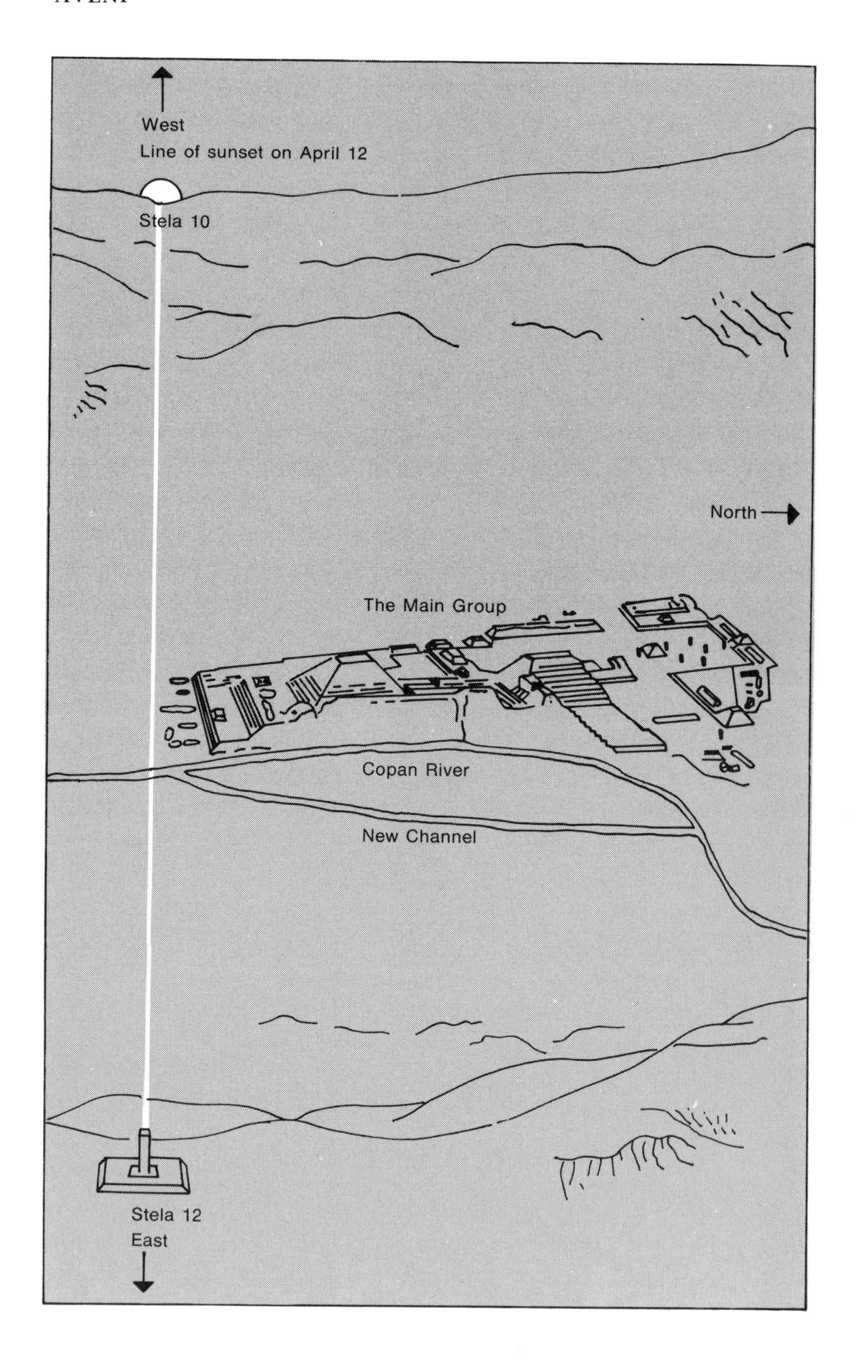
West
Line of sunset on April 12
Stela 10
North
The Main Group
Copan River
New Channel
Stela 12
East

Figure 6
The Temple of the god Venus at Copan. It bears that name because of sculptures depicting Venus symbols that are found in the doorway. A product of the Maya Classical Period (about 700 A.D.), it possesses a window facing parallel to the Copan baseline, and was probably intended to fix the same alignment.

but they developed even more sophisticated schemes than those already mentioned. At Copan, one of the most enchanting Mayan ruins, astronomer-priests carefully marked the sunset of the day the farmers were to start their agricultural calendar: they erected a pair of carved stelae at opposite ends of a seven-kilometer baseline (figures 5 and 6). The people of Copan believed that when this annual solar event occurred, it signaled the gods' permission to the people to cut and burn the bush left over from the previous year's harvest. Only then would all be in order for the planting ahead. Solar observations also foretold that nourishment from the heaven-sent rains would soon follow. But there was more than a simple agricultural-astronomical

Figure 5
The astronomical baseline at Copan, Honduras. It stretches seven kilometers from Stela 12, a monument adorned with hieroglyphs on a slope overlooking Copan, to a hill to the west of the ruins. On a date conveniently arranged by the gods to fall within 20 days of both solstice and zenith passage, the agricultural season officially commenced. The buildings in the valley were also oriented in the sacred direction. Thus the religious center of the terrestrial world could function in harmony with the spiritual world of the heavens.

arrangement here, for the sightline also served to segment the calendar into divisions which incorporated both the tropical year and a unique ritual cycle of 260 days. This latter period, found only in Mesoamerican calendrics, was composed of a sequence of 13 numbers matched successively with 20 day-names—much like our system, which pairs seven day-names with 30 (or 31) numbered days of the month. (The numbers and day-names appear in the calendar shown in figure 7.) Now either by coincidence or by design, Copan lies at precisely the latitude where the sun will spend 260 days south of the celestial zenith; on the remaining 105 days of the year, it passes north of the zenith at noon. Some investigators have suggested that the location of Copan was deliberately fixed so that the ceremonial center would be in harmony with the religious calendar.

The Teotihuacan empire of Central Mexico dominated New World cultures for nearly a millenium. In power, breadth, and longevity it may be compared with the empires of Rome or Macedonia. At Teotihuacan the expression of a knowledge of positional astronomy through the precise geometrical arrangement of buildings and ceremonial centers reached its greatest height. As testimony, a special multi-purpose symbol possessing both spatial and temporal aspects remains carved in the floors of buildings wherever the influence of Teotihuacan was felt in Mesoamerica. The key to the precise rectangular grid system underlying the scheme of the city of Teotihuacan resides in a pair of petroglyphs (figures carved in stone), one located in the center of the city, the other visible on the slope of a hill three kilometers to the west of the ruins. The line between them lies within seven minutes of arc of a perpendicular to the north-south axis of Teotihuacan. Accordingly, archaeologists who excavated the ruins during the 1960s speculated that the original architects of Teotihuacan had a definite plan in mind when they erected their city. Once the rectangular grid structure had been sketched out, all later constructions adhered rigidly to it. The peculiar orientation of the axis of the city ($15\frac{1}{2}°$ to the east of astronomical north) suggests the Teotihuacanos utilized an astronomical theme. In fact, an extension of the petroglyph base line to the west marks the place where the Pleiades would have set during the epoch in which the city was designed. As is the case at Monte Alban, a solar-stellar functional relationship is evident. About 150 B.C., when the ceremonial center was built, the Pleiades underwent their heliacal rising at Teotihuacan on precisely the same day as the first solar zenith passage. Ethnohistorians tell us that the Pleiades are well represented in Mesoamerican star lore. For the Aztecs, in fact, they were the most important celestial grouping. Priests regularly traveled to Cerro de la Estrella, the "Hill of the Star" in Tenochtitlan (ancient Mexico City) specifically to watch that prominent little cluster transit the zenith.

The form of the petroglyph symbol turns out to be every bit as interesting as the Teotihuacan orientation problem. The design occurs in one form or another thirty times in petroglyphs at ruins found from the mountainous borderland between the United States and Mexico to the rain forest of Guatemala. It usually consists of a pair of double concentric circles, quartered by a set of rectangular axes (figure 8). The design is formed by a series of circular depressions evidently made with some sort of chisel. Sixteen examples of the symbol occur in the immediate environment of the great pyramids of Teotihuacan. They are usually found on the surfaces of large rock outcrops at sites where the observer is offered a commanding view of the landscape. One particularly impressive carving (see the photograph in figure 8) is pecked onto a flat rock on a hillside at Tepeapulco, a colony of the empire located 33 kilometers to the northeast. The axis of this design points directly toward Cerro Gordo, a large mountain blocking the view of Teotihuacan. Perched atop Cerro Gordo, we find another such petroglyph overlooking both Tepeapulco and Teotihuacan itself.

Since one symbol often lies in plain view of another, it has been suggested that these designs were part of a network used to communicate information between Teotihuacan and its colonies, perhaps by signaling with polished stone or by lighting fires. To other investigators, they are merely sun symbols. But careful inspection reveals that this curious form of rock art also harbors specific astronomical and calendrical information. In a number of cases, the rectangular axes align with the summer solstice sunrise position. Two such markers are juxtaposed on Cerro El Chapin, a plateau which straddles the Tropic of Cancer (latitude 23° 27′ N). The tropic is the only place in the northern hemisphere where the zenith-passage date would correspond identically to the summer solstice. The deliberate location of the symbols at the tropic therefore probably represented an attempt by Teotihuacan astronomers to determine if this plateau, a thou-

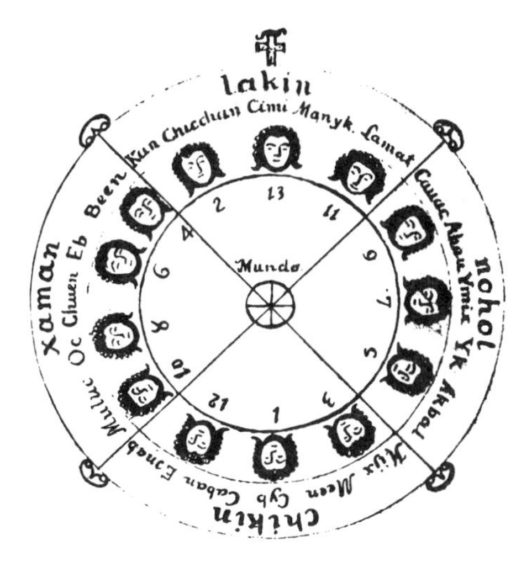

Figure 7
A calendar from the sacred book of Chilam Balam of Kaua. Like the petroglyphs that it resembles, the quartered circle embodies both spatial and temporal aspects. The four cardinal directions and the intercardinal points are attached to a fixed earth. Around the horizon which borders the diagram flow cycles of 20 named and 13 numbered days (compare with our system of weekdays with names and days of the month with numbers). The numbered days are here represented by heads.

sand kilometers north of their city, was the place where the sun turned around on its annual journey. Moreover, this pair of petroglyphs was employed to wed the solar cycle to the landscape and some of the nearby architecture by an ingenious scheme involving a double astronomical alignment. Standing on either petroglyph, an observer sees the sun rise at the June solstice over Cerro Picacho, the most prominent peak on the horizon. And at the ruins of Alta Vista, seven kilometers to the north, the observer situated in the principal structure, now called the Sun Temple, can view the sunrise at the equinoxes over the same peak. According to the archaeologist Charles Kelley, the Sun Temple was built with its corners directed toward the cardinal points of the horizon for good measure. Figure 9 illustrates this complex arrangement.

Now the Cerro El Chapin petroglyphs contain numerical as well as directional information, and again the key Mesoamerican calendric number 260 is involved; for both of the El Chapin symbols are composed of 260 depressions—72 on the axes, 84 on the inner circle, and 104 on the outer. Moreover, three-

Figure 8
Pecked-cross petroglyphs, which abound in Mexico. The symbols were carved in the floors of buildings and on rock outcrops, and seem to have been used to fix astronomical orientations; but an analysis of the design also suggests that these glyphs functioned as calendars (see figure 7). The photograph is of a glyph pecked onto a flat rock on a hillside at Tepeapulco, near Teotihuacan.

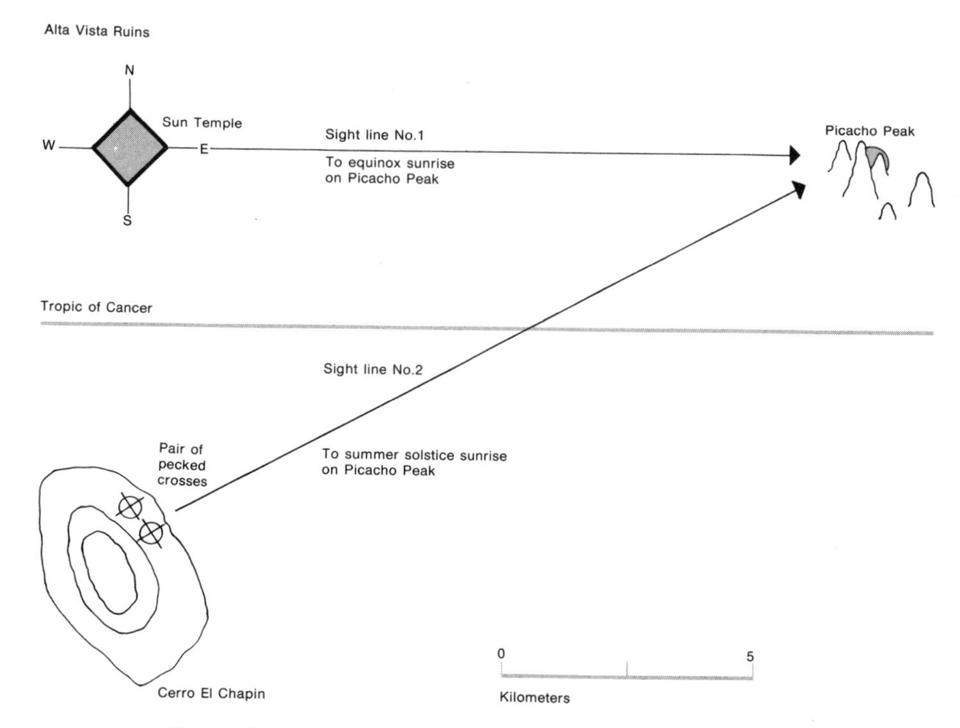

Figure 9
The Alta Vista/El Chapin double solar alignment. From a pair of petroglyphs on the plateau of Cerro El Chapin, an astronomer-priest could watch the June solstice sun rise over Cerro Picacho, the most prominent peak on the horizon. The same phenomenon, shifted to the March and September equinoxes, is visible from the Sun Temple at the Alta Vista ruins seven kilometers to the north.

quarters of the petroglyphs discovered to date display the following arrangement of points on each of their axes: 10 between the center and the inner circle, 4 between each circle, and 4 beyond the outermost circle. Thus if we move outward from the central point and include in our count the points at the intersection of an axis with either of the circles, we find 10 + 1 + 4 + 1 + 4, or 20 points, a number we might well have anticipated if we had assumed that these petroglyphs were used as mathematical devices, as it represents the number of named days in a month.

The full meaning of the symbolism of the quartered circle still eludes us, but we must face it: the creators of these designs had a multiple function in mind. These symbols served as architects' benchmarks, as astronomical orientation devices, and as calendric counters, all at once. After all, studies of calendars surviving in the pre-Hispanic codices suggest that the ancient astronomers sought to embody both spatial and temporal aspects in their cosmology. For example, the count of days and year-bearers (the names of the days of successive new years) are assigned directions in space on many of the calendars, including the one in figure 7. Moreover, the unification of space and time may be seen as well in the construction of the Copan baseline. If such a conceptualization of the world looks strange to modern eyes, consider the unification of space and time sought by the twentieth-century physicists who advanced the theory of general relativity. Einstein's approach, though more sophisticated and abstract, evokes the same theme and strives for the same goal.

In our own Western heritage, the Parthenon turns out to have an astronomical orientation: it was aligned to face the sunrise on the birthday of Athena, the goddess to whom it was dedicated. And even as late as the seventeenth century in England, cathedrals were oriented celestially. We read in an early work of that era: "One end of every Church doth point to such Place, where the Sun did rise at the time the Foundation thereof was laid, which is the Reason why all Churches do not directly point to the East; for if the Foundation was laid in June, it pointed to the North-east, where the Sun rises at that time of the year; if it was laid in the Spring or Autumn, it was directed full East; if in Winter, South-east. . . ." Thus a careful observer could tell at what time of year the foundation was laid by which direction the building was pointed.

But these may be unusual examples. Generally, in the Old World, astronomical knowledge was recorded and transmitted through the written word. The Babylonians produced cuneiform tablets, and the ancient Greeks' love of mensuration is revealed through the medium of pen and paper. In fact, in the Greek philosophical tradition, geometry and geometrical logic grew to be ends in themselves, and naked-eye astronomy provided good subject matter on which to practice this quantitative art. Often the facts of observation were secondary to the theoretical truths already inherent in a geometrical proof. Consider, for example, the argument Aristarchus employed to determine the relative distances of the sun and moon from the earth, as diagrammed in figure 10. To arrive at his result, he relied upon the concept that the interval between first and last quarter moon exceeds that between last and first. But his casual statement that

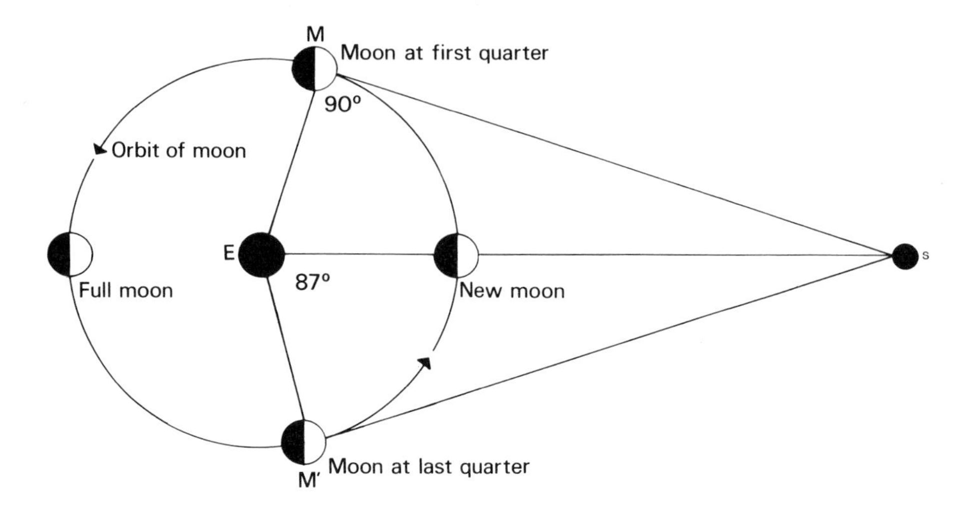

Figure 10
The Greek use of astronomy lay in the expression of their elegant system of geometry, with careful observation as a secondary consideration. Aristarchus determined the distance of the sun from the earth—SE in the drawing—as compared to that of the moon—ME (or M′E)—by solving the right triangle MES. According to him, the measure of angle M′ES—87°—was determined by the "observation" that when the moon appears to us halved, its distance from the sun is less than a quadrant by one-thirtieth of a quadrant. Thus it would take less than 14 days for the moon to pass from last quarter through new phase to first quarter and more than 15 days to complete a full lunar cycle by moving from first to last quarter. But no such difference is detectable by observation. To represent the true situation, angle MEM′ should be virtually 180° because the sun is very far away compared to the size of the lunar orbit.

he observed a genuine difference between these intervals suggests a callous disregard of the facts—or, at the least, the uncritical acceptance of an observation for the sake of an elegant and easy theorem. If one were to draw Aristarchus' diagram to proper scale, the earth-sun distance would be approximately 400 times the earth-moon distance, thus completely changing the scale and making it quite impossible to commit the beautiful proof to paper. For Aristarchus, the empirical facts of nature were subordinate to geometrical elegance.

Nonetheless, the Greeks made great strides in observational astronomy; and to appreciate the treatment they accorded the heavens we must understand that their cultural background profoundly affected the way they viewed the study of astronomy. In Greece, astronomy was not a priestly profession but instead a lay discipline which developed in a democratic society of free-thinking individuals. This made it possible to separate pure astronomical thought from other aspects of life. Unlike the Babylonians or the Maya, the Greeks were not so concerned with horoscopic problems. Thus they were released from the necessity of carefully tracking the celestial bodies: their mathematics developed into an art form. But Greek geometry, and particularly its spatial way of viewing things, gave rise to our twentieth-century models of the universe. We still depict bodies traveling in orbits, though the planets are no longer confined to the simple circles the Greeks adored.

The greatest intellectual achievements of New World astronomers are to be found not in geometry but in numerology, or so one concludes after inspecting the available source material: namely, a handful of codices which survived the sixteenth-century Spanish conquest. The Maya codices are among the most remarkable. In figure 11 are displayed portions of a table from the Dresden Codex, named after the European city in whose library it surfaced over 300 years after the conquest of Yucatan. The second and third pages tell us something about Maya knowledge of both the sun and the moon, for they are an eclipse ephemeris which served the function of warning when the face of either primary deity would suddenly vanish from view. The symbols indicated by dots and horizontal bars in the table are Maya numerals. A dot signifies one; a bar signifies five. Combinations of these symbols form numbers up to 20, the base of the Maya mathematical system. These numbers, moreover, are stacked vertically to create an ordered system of numeration with place-values which, from bottom to top, can be read in strict powers of twenty for any count other than time. It is only for the counting of the days that 360 replaces 400 as the third place-value, probably because the former lies close to the length of the tropical year. Thus the place-values in the eclipse ephemeris are 1, 20, 360 (not 400), 7,200 (not 8,000), and so on. Observe, then, that the numbers across the lower portion of the eclipse table (Row A) become, in Arabic notation, 177, 177, 177, 177, (a jump to the next page) 177, and 148; while the numbers in Row B read 6,408, 6,585, 6,762, 6,939, (jump) 7,116, and 7,264. [6,585, for example, is represented as $(5 \times 1) + (5 \times 20) + (18 \times 360)$. This number may be recognized as the Chaldean *saros*, an eclipse cycle well known in the Old World because it produces a seasonal recurrence of 35 successive eclipses of the same type.] Observe also that the numbers of Row B are obtained by adding the value of the number in Row A of the preceding column to the number in Row B directly above it. Thus 6,939 + 117 = 7,116; 7,116 + 148 = 7,264; etc. So the lower columns represent intervals, the upper cumulative totals. The number 177 may be recognized as six lunar-phase months, while 148 represents five moons.

Why did the Maya group the moons in sixes, occasionally adding a five-moon group? A study of eclipse intervals gives the answer. A lunar eclipse occurs when the full moon passes into the shadow of the earth. Conversely, a new moon passing across the solar disk produces an eclipse of the sun. Neither phenomenon can take place unless earth, moon, and sun lie nearly along a straight line. Now such a period of vulnerability to the occurrence of eclipses occurs about every 173 days—a period astronomers call the "eclipse half-year." Maya astronomers must have observed and recorded the dates of eclipses until they recognized a pattern. They then discovered that a period of six moons (177 days) lies close to this interval; an eclipse might therefore occur at every sixth moon. (Ancient Chaldean astronomers were also aware that eclipse warning was facilitated if one grouped full moons in bundles of five and six.) But since accumulated six-moon intervals creep slowly ahead of multiple eclipse half-years, the Maya also recognized that occasionally they would need to substitute a five-moon period to take up the mounting slack. The eclipse table thus seems to function as a means of warning when eclipses might occur. We can tell that the Maya must have treated such events with great seriousness, by observing the grim pictures in the lower halves of the pages.

Figure 11A-C
New World cosmology. For Ptolemy (150 A.D.), the heavens consisted of bodies orbiting the earth in divine circular paths. Though his views were modified during the Renaissance (the sun was placed at the center), the orbital concept persists in modern astronomy. The Maya, on the other hand, took the universe to consist of interlocking time cycles, two of which are captured by the maze of numbers (represented by dots and bars) in the Venus and lunar-eclipse tables of the Dresden Codex (1200 A.D.). Details are given in the text. These tables were originally published in J. Eric S. Thompson, "A Commentary on the Dresden Codex," *Memoirs of the American Philosophical Society*, **93**, 1972.

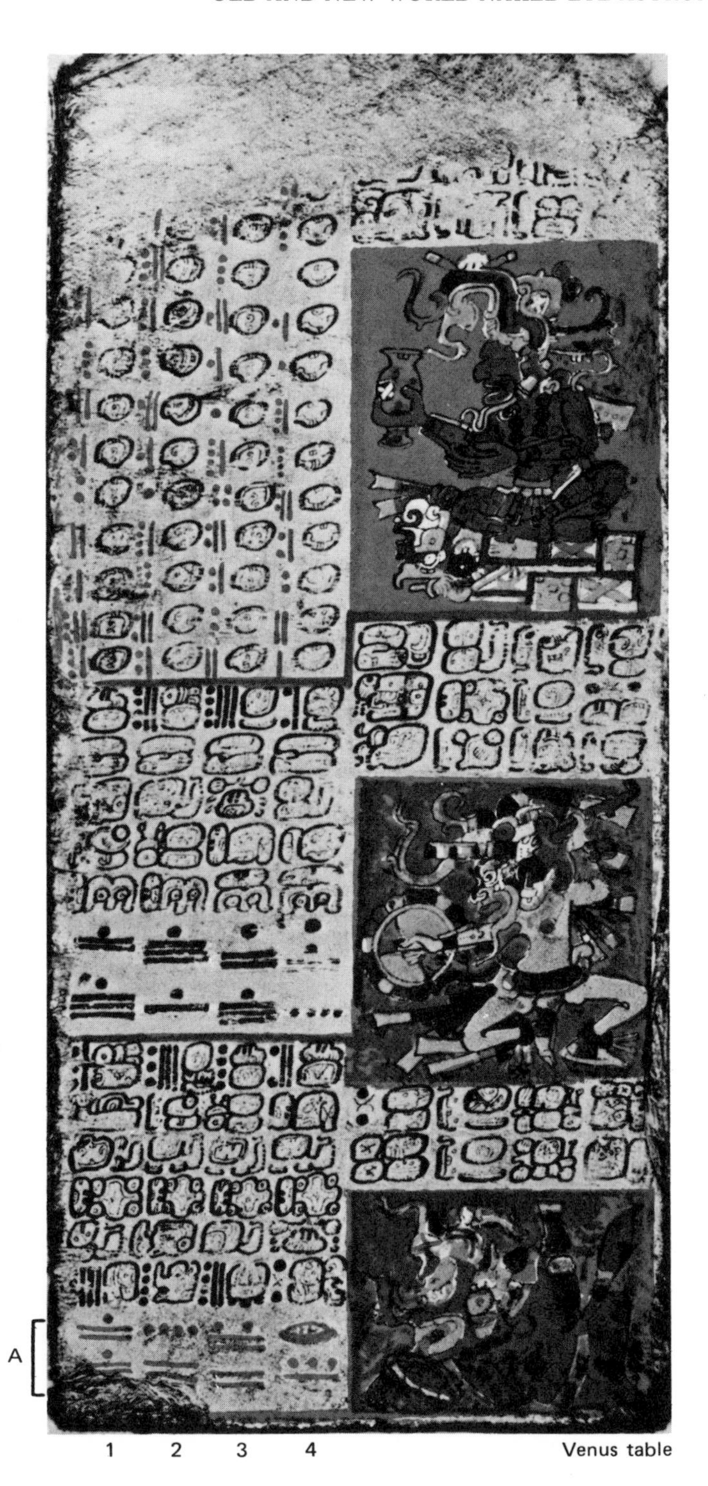

Venus table

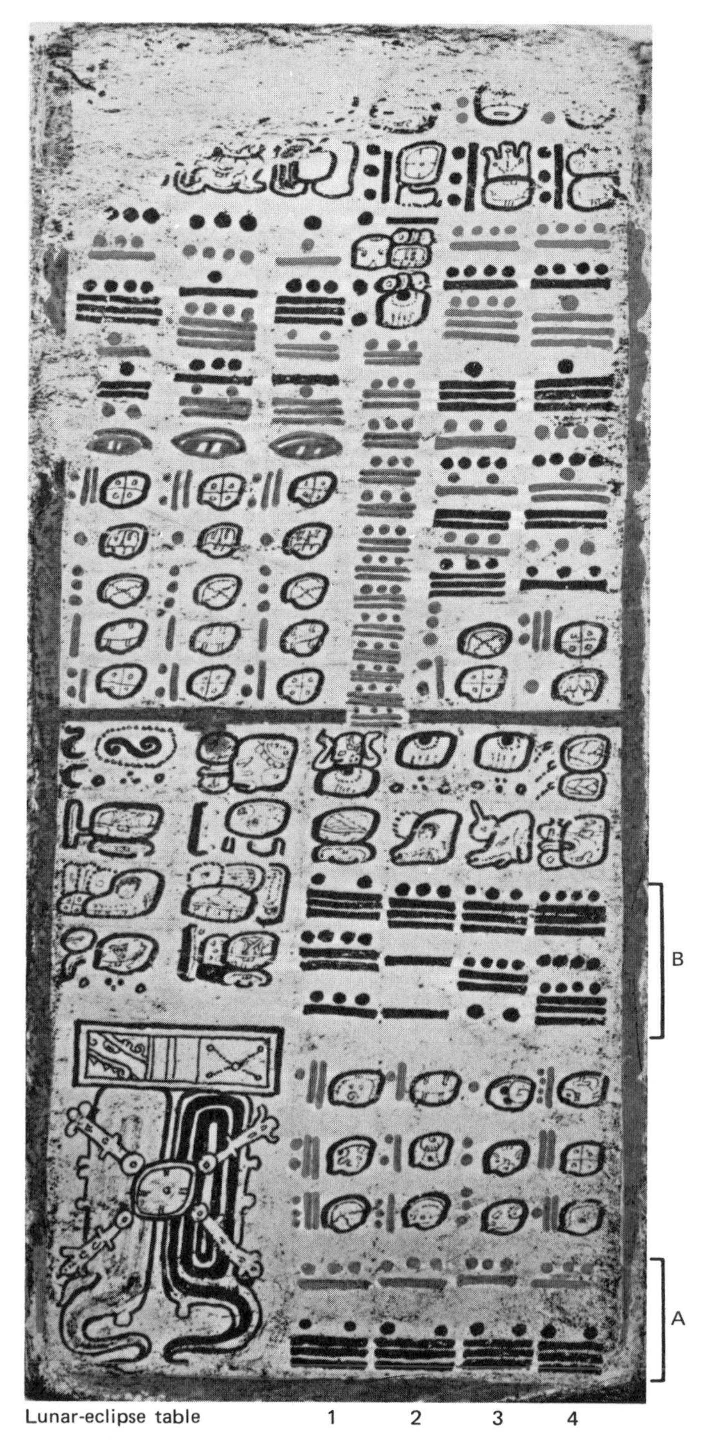

Figure 11B

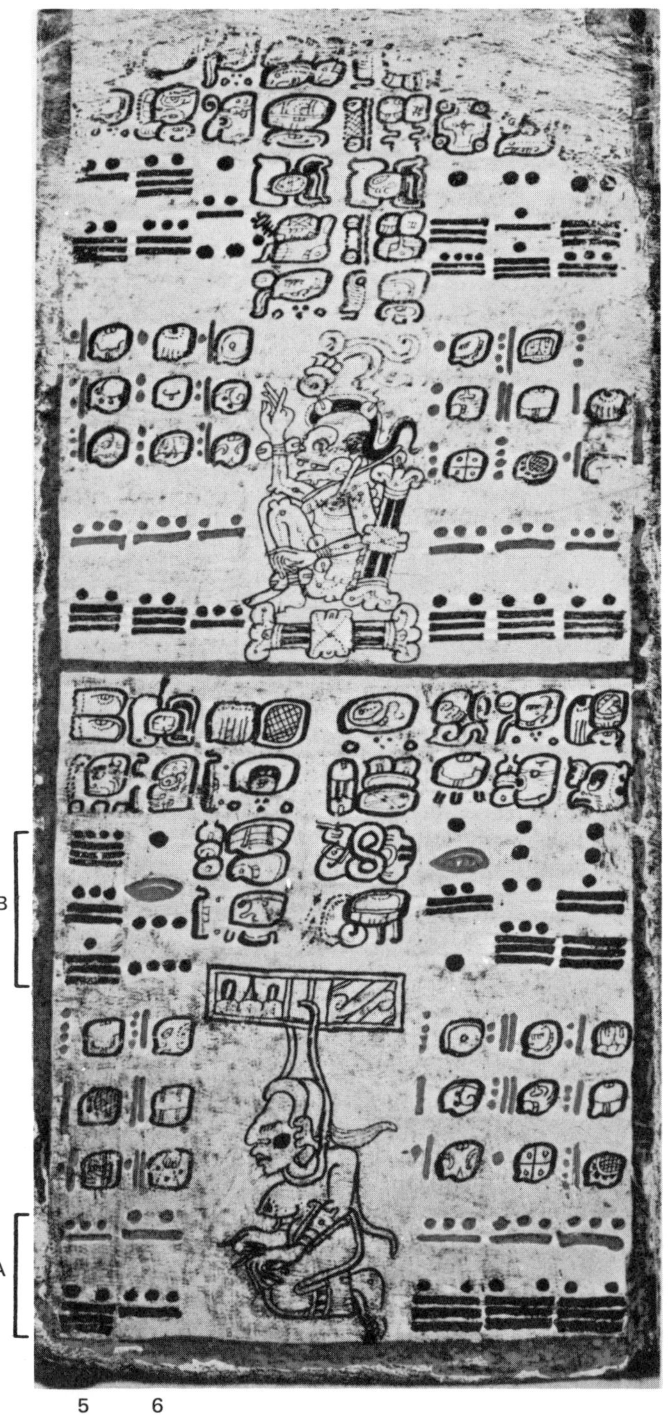

Figure 11C

If all eight pages of the lunar table were laid out we would see that it terminates with a cumulative count of 11,958 days, tallying 405 lunar months. Taking the quotient of these numbers we find that one lunar month, according to Maya determinations, is equivalent to 29.52592 days—only 7 minutes short of the modern value! Such accuracy was achieved, of course, through a long series of time-averaged observations. The lunar table also helps us to understand why these people, unlike the megalithic astronomers of Great Britain, exhibited little interest in sighting lunar alignments along the horizon. (Consult Owen Gingerich's chapter in this volume.) Their lunar coursing and eclipse prediction was accomplished quite satisfactorily simply by counting phases.

The remainder of each page of the lunar tables of the Dresden Codex consists of ritual calendar information. Maya astronomers were compelled to fashion the ephemeris to accommodate a whole multiple of their 260-day religious cycle (46 × 260 days = 11,960 days). Also, it was of the utmost importance to the Maya to be able to reenter the table after it was terminated on the same day name and number in the 260-day calendar. We often find them, like the Greeks, deliberately turning their backs on the facts of observation so that their own intellectual or religious dogmas could be upheld.

Up to this point, all of the observations and predictions we have discussed pertain to the most obvious celestial bodies in the natural environment—the sun and the moon. The planets were more elusive for ancient astronomers. Accordingly, these latter objects, so unpredictable in their courses, became the ultimate presages of the firmament. In particular, the Maya devoted to Venus, the most prominent planet in the sky, a further table in the Dresden Codex. Programatically, the Venus pages (see figure 11 for one of them) resemble the lunar. Again the lower numerals (Row A) signify intervals—this time of the four planetary stations: morning star (236 days) in Column 1 of Row A; evening star (250 days) in Column 3; and disappearance behind and in front of the sun (90 and eight days, respectively) in Columns 2 and 4. Though only one of these periods (eight days) represent a true Venus interval, their sum, 584 days, gives a very close approximation to a complete cycle of the planet.*

*As with the lunar table, the length of the Venus table was fixed by both ritualistic and astronomical considerations. Its length, 37,960 days,

Even more remarkably, the accuracy improves over the total course of the table, and, with the application of a set of corrections supplied by an adjacent table, five iterations of the ephemeris data bring the position of Venus as indicated by the Codex to within two hours of the planet's true position. But let us not allow our own cultural chauvinism to lead us to assume that a Venus "orbit" was being calculated. Time, not space, is the principal medium of expression for all the astronomy gleaned from the Maya codices.

The Venus table in the Dresden Codex is the product of post-Classic (about twelfth-century A.D.) northern Yucatan. The *Caracol,* a two-decker cylindrical building at the ruins of Chichen Itza, dates from the same era and domain. Long ago, Maya astronomers peered through the narrow horizontal slots in its turret to obtain the observations that were used to construct the Venus calendar. Today we find that sightlines measured through the Caracol's windows and in the base of the tower perfectly frame the extreme setting positions of Venus along the flat horizon. Observations of the place of disappearance of Venus in the west represent an ideal method to calculate when the planet will reappear in the east, undergoing heliacal rising after passing invisibly in front of the sun. Indeed, predictions of the heliacal rising of Venus seems to be the theme of the table. For the center pictures on each page occur adjacent to the eight-day disappearance before heliacal rise, and they depict various manifestations of the Venus god Quetzalcoatl-Kukulcan throwing spears (his dazzling rays) upon first appearance in the predawn sky. In the lower pictures we see victims impaled by the spears, symbolizing the various evil consequences which emanate from each reappearance of the god.

Having briefly inspected the Maya codices, we can see why one eminent Mayanist was moved to remark: "Let's face it, as far as the Maya were concerned, astronomy was astrology." To be sure, the regularities exhibited by the systematic recording of naked-eye observations of the moon and of Venus gave rise automatically to a scheme for predicting their motion. And the

---

is made up of 65 Venus cycles, each of 584 days. This period also coincides with a whole multiple both of 260-day cycles (146) and 365-day years (104). Furthermore, the 236- and 250-day intervals for the Venus stations correspond to 8 and $8\frac{1}{2}$ lunar months, thus suggesting that the Maya astronomers even sought to link together the Venus and lunar cycles.

system evolved into a self-correcting mechanism able to produce predictions of ever greater refinement because it was nourished by continued observation of the related events. In this sense, both the Maya and the Old World astronomers were practicing basic science. For the latter, the scientific explanation of the universe was couched in a framework of interlocking orbits, whereas the former strove for celestial harmony by appealing to the cyclical nature of time.

In order to track the planets across the sky, a naked-eye astronomer will eventually devise a zodiac. After all, the planets as well as the sun and moon all move among the stars along trajectories confined to a thin band which encircles the sky. Here, then, is a way of marking the stations or resting places of the gods by reference to different star groups or constellations which make up their heavenly roadway. Every Western culture developed a zodiac; and there exists some evidence in Native American inscriptions (figure 12) that New World people, in spite of their emphasis on horizon astronomy, did so as well.

The Chaldean cuneiform text in figure 13 is a table of motion of the planet Jupiter along a zodiacal band: it records the constellation as well as the position in degrees and minutes of arc where this second-brightest planet appeared after having executed successive complete cycles of its motion relative to the sun. Such tables disclosed minor variations in the position of Jupiter over long periods of time, which enabled Chaldean astronomers to refine their predictions for future Jupiter cycles. Like the Maya, these Old World astronomers attempted to calculate planetary periods from empirical data. But space rather than time was the medium in which their observations were couched.

Watching the fixed stars also had its rewards. Indeed, through careful observations, early Old World astronomers determined with precision one of the longest cycles in nature—the 26,000-year period of the precession of the equinoxes. This great cycle in nature is occasioned by the slow movement of the pole of rotation of the earth about the pole of its plane of revolution around the sun. The ultimate effect is a slow sliding movement of all the constellations. (Figure 1 of Harald Reiche's chapter in this volume shows the effect.) Since the zodiacal star groups were the most closely watched, observers could readily notice that their rising and setting places on the horizon were altered

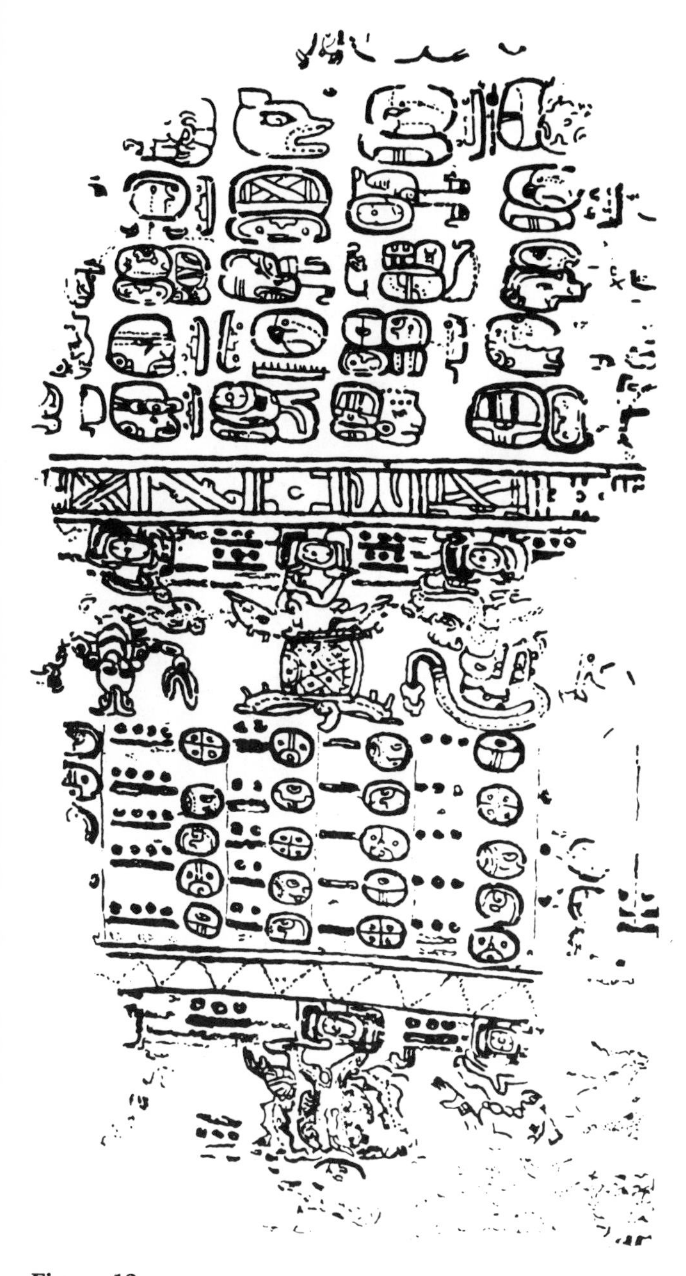

Figure 12
An apparent Maya zodiac. The illustration, taken from the Paris Codex, is a parade of 13 animals that hang from a band of sky. Though some creatures are effaced from the fifteenth-century document, a scorpion, a serpent, and a tortoise are easily visible.

Figure 13
An Old World planetary text. This Babylonian cuneiform tablet consists of stick-like figures impressed on a clay base. The numbers represent positions of the planet Jupiter along the zodiac.

through time. So, too, were the times in each year when they made their first annual appearances. There are thus at least two ways of detecting this slow transformation in the heavens. The Greek astronomer Hipparchus used the latter technique; and it has been suggested that the Mesoamericans used the former, though conclusive evidence is wanting.

In this brief review I have offered selected samples to illustrate how pretechnological peoples using only naked-eye observations of the heavens were able to build up ever more elaborate models of the universe. Such models could be used to predict the course of celestial events, but as the observations became more sophisticated, the machinery of prediction was required to yield greater accuracy. Starting with simple observations to chart the annual course of the sun, the ancient astronomers expanded their discipline to include the prediction of eclipses and the decipherment of some of nature's more elusive heavenly periodicities. Such progress is not so different from the evolution of modern science, in which an expanding technology, enabling more accurate observation, forces us to make refinements in the ways we model the behavior of nature. From our comparative study we learn that similarities between these far-flung astronomies outweigh the differences. Little wonder that diffusionists have for so long sought in ancient astronomy a support for their theories about the transoceanic migration of ideas. Yet it may be wiser to reason that since all astronomical cycles are universal, should not careful observers on both sides of the ocean arrive at similar conclusions about the heavens around them?

The motives for practicing primeval astronomy were diverse, and they are seen to be combined in all civilizations: they include chronology, agriculture, religion, astrology, civic affairs, and a curiosity about nature. It is from the early Greek philosophers, whose thoughts were filtered through the Western Renaissance, that we derive much of our present outlook. But to understand fully even our own system of thought, we must pause to look back, not merely at scientists of our own direct lineage, but rather at the astronomers of all ancient cultures. Only then can we appreciate our present-day achievements, as nested among those of the whole of human history.

For further reading:

Native American Astronomy, Anthony F. Aveni, editor, University of Texas Press, Austin, 1977.

Archaeoastronomy in Pre-Columbian America, Anthony F. Aveni, editor, University of Texas Press, Austin, 1975.

In Search of Ancient Astronomies, Edwin C. Krupp, editor, Doubleday and Co., New York, 1977.

The Mysterious Maya, by George Stuart and Gene Stuart, National Geographic Society, Special Publications Division, 1977.

A Commentary on the Dresden Codex, by J. Eric S. Thompson, *Memoirs of the American Philosophical Society,* **93** (Philadelphia, 1972).

KENNETH BRECHER

SIRIUS ENIGMAS

Sirius is the brightest star in the sky: it is about twice as bright as any other visible star. This prominence suggests that it has been well observed by astronomers and others ever since people began observing the sky, and an examination of ancient records confirms that this is so: Sirius was an important celestial object for several ancient cultures. In Babylonian cuneiform texts dating from 1000 to 500 B.C., the star is called KAK.SI.DI. The Babylonians identify it as part of a constellation which they describe as a bow and arrow; for them, Sirius was a star marking the tip of the arrow. The Chinese independently described a bow and arrow in the sky, but they used different stars for their construction. For them, Sirius was part of the image at which the arrow is shooting; and curiously, the image at which that arrow is aimed is a dog. In Western tradition, Sirius is part of the constellation Canis Major, the Big Dog. It is remarkable that the same images—dogs, bows and arrows—occur in the cosmographies of different cultures; after all, if you look at the sky, you see only points of light on a dark field. Hertha von Dechend and Giorgio de Santillana, in their book, *Hamlet's Mill,* take this as an indication that the astronomical myths of China and Mesopotamia derive from a common origin (figures 1 and 2).

Sirius was also prominent in Egyptian culture. In fact, it was perhaps one of the strongest influences upon the scientific, agricultural, and religious lives of the ancient Egyptians. The Egyptians of 3000 B.C. worshipped it for the following reason: Sirius was the star whose first heliacal rising each year—that is, its first annual rising just before the sun on the eastern horizon—occurred just prior to the flooding of the Nile. Thus the heliacal rising of Sirius each spring presaged the Egyptian agricultural cycle. We are not surprised to find that the Egyptians built temples to include alignments with Sirius (such as the small temple of Isis erected in about 700 B.C.). Even in the New World, Sirius seems to have played a prominent cultural role,

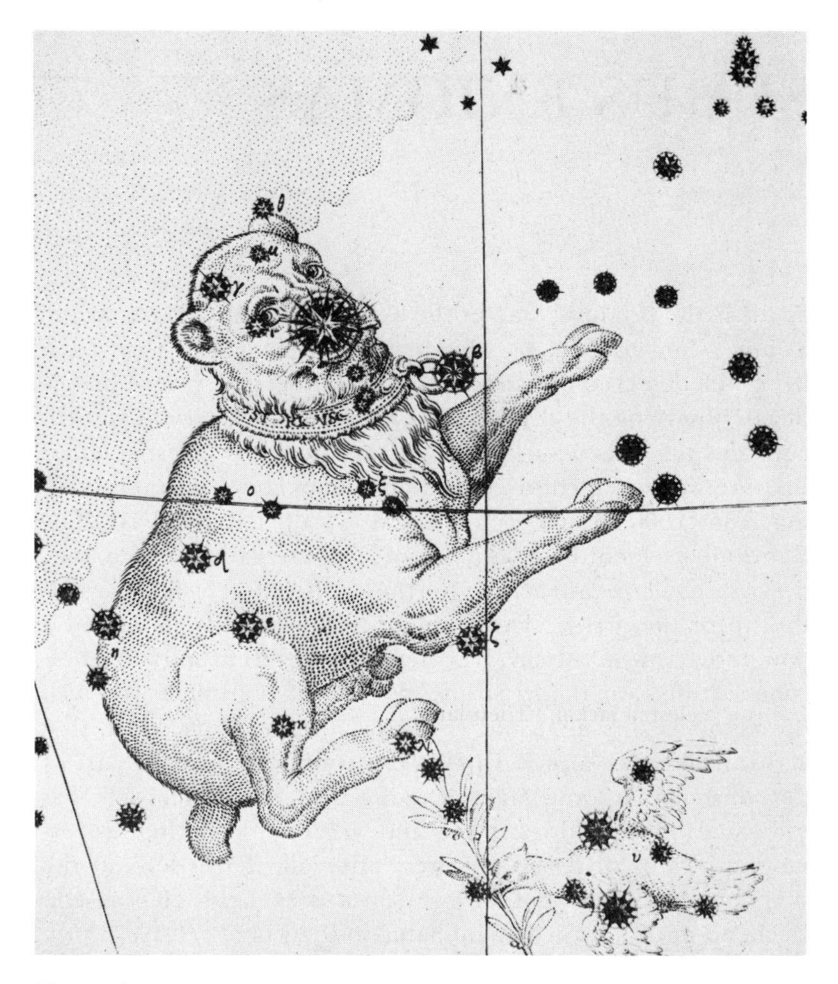

Figure 1
The celestial neighborhood of Sirius is shown in a representation from the *Uranometria,* a catalog of the sky published by Johann Bayer in 1603. Sirius itself is α Canis Majoris, the brightest star in the Great Dog, and indeed the brightest star in the sky; it appears here as the prominent star covering all of the dog's snout. The stippled region behind and to the left of Canis Major's head is part of the Milky Way. The constellation appearing at the lower right of the illustration is Columba, the dove.

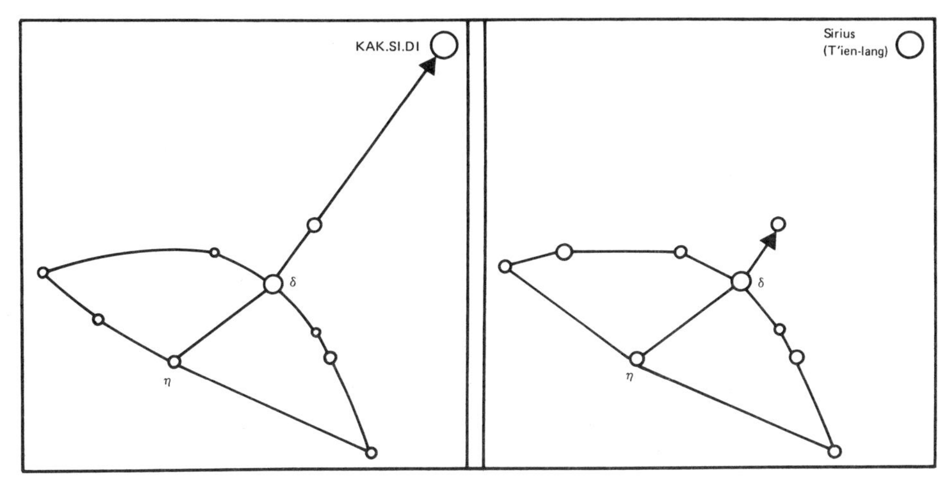

Figure 2
The stars in the vicinity of Sirius, as grouped into constellations by the Babylonians (left) and the Chinese (right). In both cultures, the stars are construed as composing a bow and arrow. For the Babylonians, however, the arrow is relatively long, and Sirius is at its tip. By contrast, the Chinese made Sirius the target at which the arrow is aimed. Indeed, the Chinese were specific about the nature of that target: they called Sirius a "celestial jackal," T'ien-lang.

being incorporated in the design of the Bighorn medicine wheel (see John A. Eddy's article earlier in this issue) and in the street layout of the ancient city of Teotihuacan in Mexico.

My own interest in Sirius derives from my hope that one can learn astrophysics from history. It has happened before. Many of us in high-energy astrophysics consider the Crab Nebula, for example, to be our Rosetta Stone: a key, not to ancient myths, but to modern physics. The Crab Nebula is the remnant of a supernova explosion, and near its center lies a pulsar, a pulsing radiation source reliably believed to be a rotating neutron star. Much of what we know about the astrophysical origin of cosmic rays, synchrotron radiation, and heavy elements derives from our knowledge about the Crab Nebula, and much of what we can deduce from observations of the Crab Nebula is aided by knowing that the supernova explosion that created it was recorded in 1054 A.D. by Chinese astronomers, and perhaps by Southwest American Indians (see John C. Brandt's chapter in this volume). Now Sirius, as I will show later in this article, might be called the Rosetta Stone of normal stellar astronomy—the study of rather normal stars that have never exploded but have gently shriveled up and are fading away. Accordingly, anything that ancient civilizations can teach us about Sirius is likely to be useful in understanding the late stages of stellar evolution.

Another motivation for studying the ancient references to Sirius—or for studying archaeoastronomy in general—is essentially a negative one: over the past few years, writers such as Erich von Däniken have achieved a great deal of popularity by suggesting that the ancients cannot possibly have known as much as they apparently did about astronomy, unless they had help from aliens in UFOs. Let me give a hypothetical example of the specious sort of reasoning one might apply to Sirius. There is a number which shows up repeatedly in ancient numerology called the Golden Mean. For modern mathematicians, it is the solution to the equation $x - 1 = 1/x$. Alternatively, it is derived by constructing a Fibonacci series—a progression of numbers beginning with 1, in which each number is the sum of the preceding two. Thus the series is 1, 1, 2, 3, 5, 8, and so on. As the numbers of the series approach infinity, the ratio of the last two numbers approaches the Golden Mean. Whatever its derivation, the Golden Mean is $(1 + \sqrt{5})/2$, or about 1.61. Various shells in nature are constructed on the basis of the Golden Mean; the spiralling shell of the chambered nautilus, for instance, conforms to the law generating the Fibonacci series.

The Greeks and Egyptians considered the Golden Mean a mystical quantity. They also considered geometrical representations of the Mean to be pleasing to the eye, for the sides of the Parthenon are constructed on the ratio of 1.61 to 1. The ratio of the base to the height of some Egyptian pyramids is said to be the same. Now I have already said that the Egyptians were very interested in Sirius. I have not yet said that Sirius is actually a binary star system in which the visible star is accompanied by an essentially invisible companion—a second star which cannot be seen without telescopes (figure 3). The eccentricity of its orbit (a mathematical term denoting its deviation from circularity) happens to be about equal to the inverse of the Golden Mean.

But is the reader inclined by these circumstances to suppose that the Egyptians built their pyramids in conformity with the orbital eccentricity of the invisible companion to Sirius? And did the Egyptians, lacking powerful telescopes, receive this knowledge from creatures of another world? I don't think so. The earthly sources of the Golden Mean seem to be more probable. Of course, the derivation of the Golden Mean in any

Figure 3
Sirius as it appears in a powerful telescope. The bright, six-pointed image at the center of the photograph is the very same Sirius that can be seen with the unaided eye. Actually, however, this object is Sirius A, for "Sirius" has been known since 1862 to be a binary star system. The companion star, Sirius B, is too dim to be seen with the unaided eye; here it appears below and to the right of Sirius A. The other points of light in the photograph, strung out in a horizontal line at either side of Sirius A, are all artifacts. They were generated by placing a fine-wire grating in front of the telescope's objective lens, and are used for positional calculations that cannot be made with the bright, fuzzy image of Sirius A itself. Finally, the six points radiating from Sirius A are artifacts as well. They were generated by a hexagonal diaphragm placed in front of the objective lens, and serve to channel the light of Sirius A in such a way as to leave areas of blackness in which Sirius B and the innermost two artifactual points can be seen. The photograph was taken by Irving W. Lindenblad at the US Naval Observatory.

fashion requires ingenuity. And thus our study of ancient astronomy cannot help but increase our respect for the intelligence of our predecessors—even if they *didn't* know about the companion to Sirius.

With all this as background, I propose to discuss two mysteries surrounding the brightest star in the sky: one historical, the other mythological. The first of these mysteries is the following: beginning about 1000 B.C. and continuing through about 200 A.D., Sirius was said to be red by every writer who wrote about the matter (table 1). Now if you go outside and look at the brightest star in the sky you will find that Sirius is white, or perhaps blue-white. It surely isn't red like Mars. Nor is it red

Table 1
Ancient references to the color of Sirius. Explicit namings of red span a millennium, from a Babylonian cuneiform text to the Greek astronomer Ptolemy. All seem agreed that the Dog Star is red, but this is distressing, for the star is blue-white today, and modern theories of stellar evolution cannot account for it being a different color in historical times.

Source	Reference
Egyptian glyphs (ca. 2800 B.C.)	*"Sothis [Sirius], herald of the new year and the flood";* no explicit color reference.
Babylonian cuneiform text (ca. 700 B.C.)	A star called KAK.SI.DI "shines like copper."
Aratus (ca. 270 B.C.)	*ποικίλος* (*poikilos,* or "colored").
Cicero (ca. 50 B.C.)	*Rutilo com lumine* ("with ruddy light").
Horace (ca. 10 B.C.)	*Rubra canicula* ("red dog").
Seneca (ca. 25 A.D.)	*Acrior sit Caniculae rubor, Martis remissior, Jovis nullus.* ("The redness of the Dog Star is deeper, that of Mars milder, that of Jupiter nothing at all.")
Ptolemy (ca. 150 A.D.)	*ὑπόκιρρος* (*hipokeros,* or red, coppery, yellowish) used for Aldebaran, Betelgeuse, Arcturus, Antares, Pollux (all currently red stars), and Sirius.
Al Sufi (ca. 980 A.D.)	Red stars include all those named by Ptolemy except Sirius.
Ulugh Beigh (ca. 1450 A.D.)	Red stars include all those named by Ptolemy except Sirius.

like the stars that actually *do* look reddish: Aldebaran, Betelgeuse, Antares, Arcturus, and Pollux. Perhaps, you might say, there is no puzzle in this: Sirius was red then and blue-white now. The problem is that, so far as any astrophysicist can understand, there is no theory of stellar evolution that will make Sirius change from red to blue-white in 2,000 years.

With fear, therefore, for the tenability of accepted theories of stellar evolution, let us examine the ancient evidence. In a Babylonian cuneiform text dating from about 700 B.C., one finds a reference to the star "KAK.SI.DI. rising in late autumn and shining like copper"—or so it says in the translation I have read. I myself do not know how this translation comes about. I do know that language in general is riddled with ambiguity; for instance, the Russian work *krasnoy* can mean "red" or "magnificent." Thus you cannot know the meaning of the word unless you know the context in which it appears. Accordingly, I'm not sure what to make of a star that "shines like copper." The phrase might indicate that Sirius had the color of copper—or perhaps the brilliancy of polished copper.

What did the Egyptians have to say about Sirius? Unfortunately, the Egyptians were inclined more to astrology than to astronomy; there is little of any sort to be gleaned about quantitative stellar astronomy from their writings. In particular, although the Egyptians watched Sirius with intense interest (it is the only star explicitly named in any of their records), they have left no surviving references to its magnitude, its color, its changing position, or any other physical property.

In the time of the Romans, the astronomical record is more extensive. Sirius is sometimes called the Dog Star, and sometimes it is called "rubra-canicula," the red dog. I don't think there can be any ambiguity in the translation of that latter expression. In addition, there are the references to Sirius by Seneca, a sharp observer who seems to have interpreted astronomical observations more correctly than Aristotle; he knew, for example, that comets were not an atmospheric phenomenon, and explicitly wrote that they must orbit the sun. In reference to Sirius, Seneca reported that "the redness of the Dog Star is deeper, that of Mars milder, that of Jupiter nothing at all." The translation seems unambiguous.

Then there are the writings of many other Roman and Greek authors, Pliny, Cicero, Horace, Ovid, Hesiod, Homer, and Virgil among them: they all call Sirius red in one way or another. But the evidence that makes the strongest impression derives from Ptolemy, perhaps the greatest astronomer of antiquity. In about 150 A.D. Ptolemy recorded most of the Greeks' knowledge of astronomy in his book *He Mathematike Syntaxis,* later translated by the Arabs under the name *Ho Megas Astronomos* ("The Great Astronomer"). In the ninth century the Arabs further corrupted the title, changing the Greek "Megas" to "Megiste," adding the article *al,* and finally transforming it to *Almagest,* the name under which the book is known to us today. The *Almagest* includes a set of tables listing over a thousand stars, most of which were first studied and catalogued by Hipparchus in about the year 133 B.C. Several characteristics of the stars are given, and among these characteristics is color. In particular, six bright stars are called red: Aldebaran, Betalgeuse, Antares, Arcturus, Pollux—and Sirius. The first five of these are in fact the brightest red stars in the sky; all can be seen quite easily today, and all are reddish to the eye; they all fall into the category of stars known to modern astronomers as red giants.

But what about the sixth? Could the astute and clear-minded Ptolemy have made a mistake? And if not, why should Sirius be called red 2,000 years ago? One possible reason is that some prestigious ancient writer made a dreadful mistake about the color and all subsequent scholars followed suit. After all, some happenings of that sort are famous: it appears, for instance, that no one of authority counted the teeth in a horse's mouth after Aristotle gave the incorrect number. Still, I don't think that anything similar could have happened for the color of Sirius because the references in question are Greek, Roman, and Babylonian: that is to say, they derive from a long period of history and many different cultures. It is hard for me to believe that the mistake could be perpetuated very long, with Sirius so prominent in the sky.

Another possibility is that the ancient observers looked at Sirius only when it was rising and near the horizon. Light rays reaching an observer at such an angle undergo a fair amount of reddening, and so Sirius presumably would have looked red until it had risen higher in the sky. Now if that was all there were to this explanation, it would be a pretty feeble one. After all, why should the ancients have looked at Sirius only when it was near the horizon? Moreover, at the latitude of Alexandria, where Ptolemy wrote, the star α Centauri never rises more than a few degrees above the horizon. Why doesn't the *Almagest* describe α Centauri as red?

There is a clever idea that will obviate many of these problems; it was mentioned to me by Philip Morrison but I subsequently found that the 19th-century Italian astronomer Giovanni Schiaparelli had thought of the same thing a hundred years earlier. The idea is the following: There is in fact only one star in the sky that is bright enough to be seen as a reddened star on the horizon. That star is Sirius. (In the words of Tennyson, ". . . the fiery Sirius alters hue/And bickers into red and emerald. . ." The Arab poet Ibn Alraqqa wrote: "I recognize Sirius shining red, whilst the morning is becoming white./The night, fading away, has risen and left him./The night is not afraid to lose him, since he follows her.") Since the Egyptians were interested in Sirius at its heliacal rising, it is therefore natural that they might have called it red. That explanation, of course, fails to cover the references to the redness of Sirius made by other cultures, unless the Roman tradition appropriated the redness from the Egyptians. But then there are the Babylonian references to be accounted for.

Still another possibility is mistranslation of the word used by Ptolemy—the word *hipokeros* (νποκίρρος). As I noted earlier, that isn't likely. Still, an argument could be made that we really don't have original copies of the ancient books. In particular, we don't have a 2,000-year-old copy of the *Almagest.* We have instead a version that was translated into Arabic around the ninth century A.D. and then translated back into Greek. It is probably riddled with corruptions. A typical example in fact turns out to be of importance for my topic: at the end of the *Almagest* is a summary of what Ptolemy saw in the sky. One of the items, according to the Arab translation, is *five* red stars. Now if you consult the bulk of the book and make your own compilation of stars said to be red, you will find a total of six—the six stars I named earlier, with Sirius among them. Is it fair to deduce from this that some scribe mistakenly wrote down Sirius as being red in the bulk of the book, but in the summary wrote the correct total of five? There is a second surviving version of Ptolemy, a version that has never been translated from the Greek, but only transcribed over and over again. The copies that we have of this second version date from the fourteenth and fifteenth centuries. And these manuscript copies, translated into English, report in the summary that there are five *nebulous* stars—whatever that means.

Let me elaborate briefly on why the redness of Sirius in historical times, if true, would be a puzzle for astrophysicists.

Sirius, as I mentioned earlier, is in fact two stars. The bright star that we see—and that the Egyptians presumably saw—in the sky is Sirius A. It has a magnitude of −1.5, and is regarded by astronomers as a perfectly normal (A-type) star: nothing much has happened to it, we think, in the last 1 to 10 billion years. The companion star, Sirius B, is ten magnitudes, or a factor of 10,000 times, dimmer. It cannot be seen without a telescope.

Sirius B is not at all a normal star. It is a white dwarf: a star of about the same mass as the sun, but with that mass so condensed that the radius of the dwarf is 100 times smaller than the radius of the sun (about equal to the radius of the earth). The dwarf is supported against further collapse by a pressure that can be accounted for only by quantum mechanics, and not by classical physics. How does such a star arise? Stars in general are thought to begin their lives by the nuclear combustion of hydrogen. They pass from that stage to a stage in which they burn helium, and in doing so they expand into what are called red giants. Eventually the giant's atmosphere of helium is blown off or otherwise consumed. What remains is a white dwarf (figure 4). Now the lifespan of a red giant is thought to be tens of millions of years, and the blow-off time during which a red giant becomes a white dwarf and cools to the presumed temperature of Sirius is also thought to be a truly astronomical time: say a million years. Accordingly, if Sirius looked red to the ancients because Sirius B, a white dwarf now, was a red giant only 2,000 years ago, all theories of stellar evolution would have to change.

How then does an astrophysicist approach the problem of why Sirius was said to be red? Here are the possibilities that come to mind: first of all, the astrophysicist might conclude that it simply *wasn't* red; it was blue-white. After all, that's the most acceptable astrophysical explanation.

In the second place, it is possible that a dust cloud passed in front of Sirius roughly 2,000 years ago. There are in fact dust clouds in interstellar space and they do filter light in such a way that objects behind them appear to be redder. Of course the

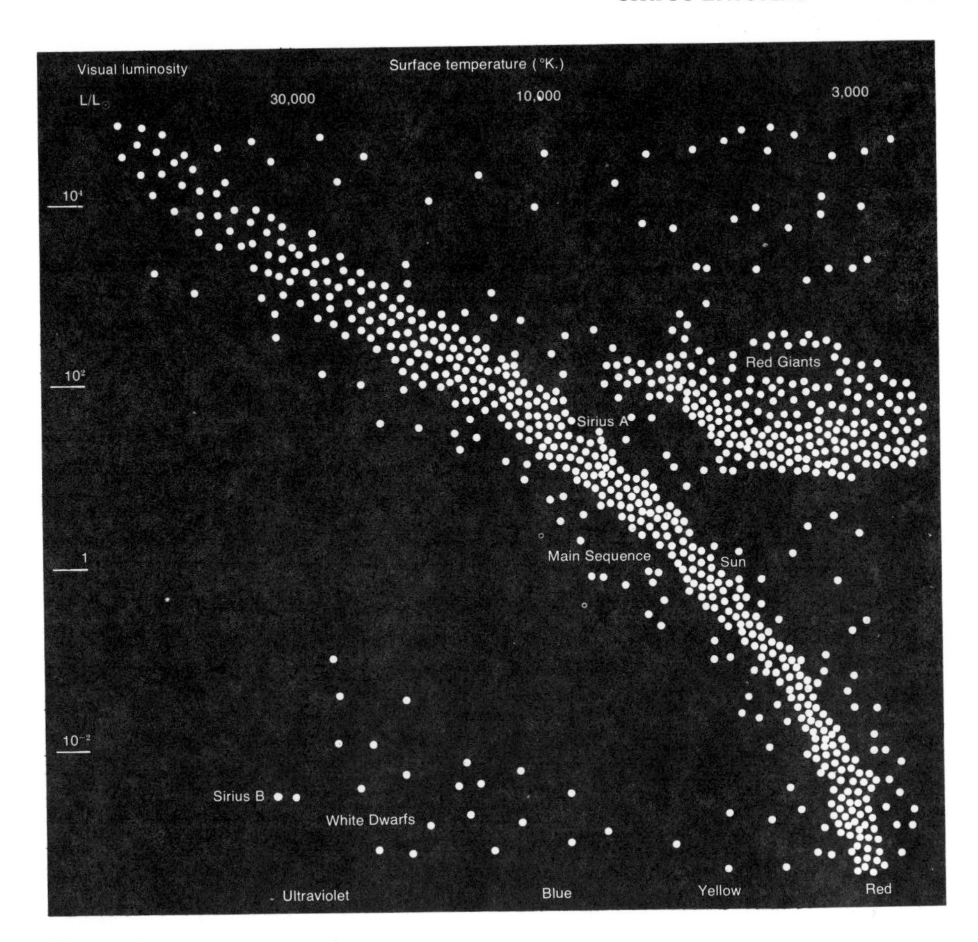

Figure 4
The nature of stars is displayed on a Hertzsprung-Russell diagram, in which a star's surface temperature is plotted against its visual luminosity. (The luminosity of our sun is taken as unity.) A so-called main sequence runs from the upper left to the lower right of the chart; massive stars tend to be more luminous and have higher surface temperatures, placing them toward the left of the sequence. After a star consumes its hydrogen, it leaves the main sequence and expands into a red giant, a hundred times as bright as the sun but redder in color. The star now burns helium. Eventually, though, the red giant phase ends: the extended outer envelope of the star is blown off or otherwise lost, leaving behind a hot, dense star called a white dwarf. All of these events are thought to occur on a time scale measured in millions of years. Sirius A and Sirius B are placed on the diagram in accordance with their observed colors and visual luminosities. Sirius A, a main sequence star, is about ten thousand times brighter than Sirius B, a white dwarf.

objects would also appear to be dimmer, but Sirius is so bright that even if it lost two or three magnitudes, it would still be a first-magnitude star. One wonders, though, where the dust cloud is now.

A third possibility is that Sirius B was a red giant a million years ago—say in pre-Neanderthal times. I think many astrophysicists would reluctantly allow that. If so, one could claim (with every anthropologist in the world convulsed by laughter) that the Neanderthals observed the sky, recorded the fact that Sirius was red, and passed down that fact in a tradition that lasted a million years. It doesn't seem very likely.

A fourth possibility is that Sirius B was indeed a red giant 2,000 years ago. If so, its magnitude would probably have been −3 to −8. It would have been a spectacular star, visible even in the daytime, and that would surely explain why so many different cultures were interested in it. On the other hand, I repeat that one simply cannot get the transformation from red giant to white dwarf to work in 2,000 years—not even with all the latest possibilities envisaged for subatomic physics by theoretical physicists. In particular, a 2,000-year transformation would seemingly leave the surface temperature of Sirius B at about a million degrees Kelvin or more—far higher than its recently-measured temperature of 30,000 °K.

A fifth possibility is that mass transferred from Sirius A to Sirius B. This happens, for example, in novae—two-star systems, it is now thought, in which a star seemingly explodes over and over again because it keeps receiving mass. To be sure, it is highly unlikely that mass transfer could have occurred in the Sirius system, for those stars are relatively far apart as binary star systems go. Yet if it had in fact occurred, Sirius B might temporarily have become a red giant of sorts, or actually what I would call a slow nova. Examples are known of stars that have made similar detours from normal stellar evolution, such as F G Sagittae or η Carina.

Sirius has played a prominent role in stellar astronomy and physics over the past 200 years. In 1710, for example, Edmund Halley compared the various star positions he had measured with those measured 1,500 years earlier by Ptolemy and decided that Sirius, Arcturus, and Aldebaran were in the wrong places in the sky. They were wrong by about a degree—twice the diameter of the sun. Halley could hardly think that Ptolemy could have made so big an error. Therefore, he concluded, the stars must move amongst each other.

A much more subtle finding came in 1836, when Friedrich Bessel made a careful study of the motion of Sirius. (Bessel is best known as the mathematician who developed Bessel functions, but he did so as an aid to his astronomical researches.) Bessel determined that Sirius was moving against the background of stars at the rate of about one arc-second per year, but that it was wriggling sinusoidally as well, with a maximum amplitude of 11 arc-seconds and a period of about 50 years. From this he concluded that an invisible companion star was (and is) perturbing the visible star's trajectory. About this star he wrote a famous letter to Alexander Von Humboldt—a letter relevant, no doubt, to all people who espouse the cause of black holes today. "I adhere," Bessel wrote, "to the conviction that [the stars] Procyon and Sirius form real binary star systems consisting of a visible and an invisible star. There is no reason to suppose luminosity is an essential quality of cosmical bodies. And visibility of countless stars is no argument against the invisibility of countless others." In short, the absence of evidence, claimed Bessel, is not the evidence of absence. He started what he called the astronomy of the invisible.

About 20 years later, Alvan Clark, the greatest 19th century telescope maker, collaborated with his son on the construction of an 18½-inch refractor, at that time the largest telescope in the world (figure 5). They tried it out at their workshop in Cambridgeport, Massachusetts, on the bank of the Charles River where the Boston University Bridge now stands; and they trained it on Sirius as it set over Beacon Hill. A second star promptly appeared in the field of view right next to Sirius—a very dim star, situated in exactly the position that had been predicted from Bessel's calculations. In this way Sirius B was discovered in 1862.

I have tried without success to find out what sort of attention this discovery received at the time. It should have received a great deal, for Sirius B was very dim and yet, using Newtonian mechanics, one could calculate that it perturbed Sirius A as if it had the mass of the sun. The conclusion at the time seems to have been that Sirius B must be very cool. That, after all, would explain the dimness—and it did, until 1910, when it was discovered that certain dim stars have spectra indicating that they are hot. Now if a star is both hot *and* dim, one can conclude only

Figure 5
Alvan Clark's workshop ("Clark and Sons' Observatory") on Henry Street in Cambridgeport, Massachusetts. In the foreground are a telescope mount and tube which were used for testing objective lenses; the Clarks made the largest lenses in the world. While testing an eighteen-and-a-half-inch objective—indeed the largest in the world at that time—the Clarks discovered Sirius B, the nearly invisible companion star to Sirius A. The lens itself was to have gone to the University of Mississippi, but the Civil War intervened. Harvard College then wished to purchase it. The ready cash of the University of Chicago won out. The Chicagoans paid $11,187 in 1863.

that it is small. In 1914, Walter Adams, working at Mount Wilson Observatory, managed to measure the surface temperature of Sirius B. It turned out—incorrectly, in retrospect—to be about 8,000 °K. It followed from this measurement that the radius of Sirius B, to account for the dimness of the star, would have to be a hundred times smaller than the radius of the sun; and that in turn meant that for the star's mass to be equal to that of the sun, its density would have to be a million times greater—its mean density would have to be a million grams per cubic centimeter. This was astonishing. Sir Arthur Eddington's popular books on astronomy, written in the late 1920s, confirm that nobody believed there could be anything in the universe anywhere near so dense. I have found references in the 1920s to Sirius B in *Le Monde,* The *New York Times,* and *Scientific American.* It was the black hole of its day.

Now at the same time as Sirius B was creating this furor, the theory of general relativity was being developed by Albert Einstein. In 1919, Eddington, who was actually a theorist, nevertheless went on the famous expedition in which the bending of starlight by the sun during a solar eclipse was measured. It was found to conform well to the prediction made by Einstein's theory. Einstein promptly became a cult hero—about as big as the Beatles, judging from the press coverage at the time. In 1925, Eddington suggested that Adams (the Mount Wilson astronomer) ought to measure the shift in the spectral lines of light leaving the surface of Sirius B. For if that star were truly as dense as was supposed, its gravitational field would be prodigious. And according to general relativity, light has a hard time escaping from a strong gravitational field—a hard time that finds expression in a shift of spectral lines toward the red. Adams did measure the red shift, and found it to be in good agreement with the calculation made by Eddington. Moreover, it agreed with the tiny radius attributed to the star. Here then was a true mystery for astrophysicists: Sirius B did in fact have an unbelievable density. How was this possible? Nearly a decade after the mystery arose, it was finally solved: R. H. Fowler, in one of the first applications of quantum mechanics to a macroscopic system, showed in 1925 that the new physics could indeed allow a star to exist at a density of a million grams per cubic centimeter by providing so-called "degeneracy" pressure—without analog in classical physics—to support it against further gravitational collapse.

I come now to the second engima surrounding Sirius—a mythological enigma. It concerns a tribe in Africa called the Dogon, which lives in what is now Mali but was called the French West Sudan 50 years ago (see figure 6). In any event, the tribe lives near the city of Timbuktu, and directly on an African trade route from Egypt to West Africa. European contact with the Dogon seems to have been limited to the arrival of an occasional missionary, at least until the early years of the 20th century. Beginning in the 1930s, however, two French anthropologists, Marcel Griaule and Germaine Dieterlen, lived with the tribe for over 20 years. As time went on they gained the tribe's confidence. Indeed, after about 20 years Griaule was so highly regarded that the tribe decided to tell him their deepest secrets. Accordingly, the tribe picked four old chiefs to tell Griaule about the Dogon's cosmology—their knowledge of the stars. The four informants were Innekouzou Dolo, a woman between 65 and 70 years of age, priestess of Amma and soothsayer, who spoke the Sanga language; Manda D'Orosongo, 45-year-old priest of the Binou Manda, who spoke the language Wazouba; Yébéné, 50-year-old priest of the Binou Yébéné, who spoke Sanga; and Ongnonlou Dola, between 60 and 65 years old, patriarch of the village of Go, who also spoke Sanga.

What did the Dogon know about the sky? In the first place the informants told Griaule that Sirius is at the center of their mythology. And then, drawing on the ground with a stick, they showed him that Sirius was not at all a single star; it was two stars. Indeed, to quote Griaule and Dieterlen: "[the visible Sirius] is one of the foci of the orbit of a tiny star called Digitaria, potolo, or the star of the Yourougou, Yurugu tolo, which plays a crucial role, and which, unaided as it were, hogs the attention of male initiates." Note (from figure 7) that the Dogon placed Sirius A at the focus of an ellipse, not at the center of a circle. The ellipse represents the orbit of Digitaria, the invisible companion star; and that companion orbits Sirius A in a period which, counted twice by the Dogon (for a reason having to do with the primacy of a concept of twin-ness in the Dogon worldview), is 100 years. Further still, Digitaria was said by the Dogon to be the beginning and end of all things; and because of this role the star is considered to be the smallest thing in the sky—and also the heaviest. Digitaria, reported Griaule and Dieterlen, "consists of a metal called sagala, which is a little brighter than

Figure 6A-D
Four views of the Dogon civilization. Photographs A and B were taken by Jay M. Pasachoff, an astronomer at Williams College, during a visit to West Africa occasioned by a total solar eclipse. The hogon (priest) and his temple (C and D) were photographed by Kenneth Strzepek, a graduate student in MIT's Civil Engineering Department, in the course of a water-conservation project in which he participated.

Figure 6B

Figure 6C

Figure 6D

iron and so heavy 'that all earthly beings combined cannot lift it.'"

Griaule and Dieterlen were also told that the earth rotates around the sun, that Jupiter has four satellites, that Saturn has rings not like the ring occasionally seen around the moon, that the planets all move around the sun, and that they always do so in elliptical rather than circular orbits! The problem for us, therefore, is how the Dogon could have known a host of astronomical facts, all of which are invisible to the unaided eye. In particular, how could they have known about the existence of Sirius B? How could they have known of its incredible density? its elliptical orbit? its 50-year period? They have no business knowing any of this.

The most obvious possibility is the dreadful thought that it's all just a fake. When Griaule died, he was so much a member of the tribe that about a quarter of a million Dogon gathered and held a state funeral. He was the only outsider ever so honored. Maybe this was his last joke. I am told, however, by Hertha von Dechend that both Griaule and Dieterlen performed scrupulously honest research; they weren't even flamboyant people, so one cannot accuse them of trying to achieve notoriety with their spectacular story. The journal article in which they report their findings is certainly sober enough.

A second obvious possibility is that somebody told the Dogon

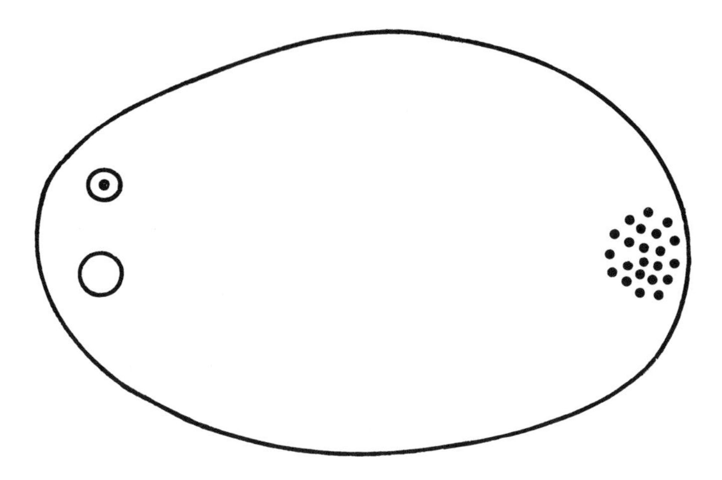

Figure 7
The Dogon's understanding of the binary star system Sirius, as drawn on the ground with a stick by elders of the Dogon tribe of the Mali Republic. The open circle at the left is the star known to astronomers of our own civilization as Sirius A; it is a prominent star, easily visible to the unaided eye. Above it, and represented as a dot within a circle, is a star not visible to the naked eye that the Dogon call Digitaria; it is shown, according to the Dogon, in its position nearest the visible Sirius. The cluster of dots near the right edge of the ellipse, again according to the Dogon, is Digitaria at its greatest excursion away from its visible companion. Digitaria is evidently depicted as a set of dots in this latter place in order to represent a "twinkling" that the Dogon attribute to the star at that point in its orbit.

all about Sirius before Griaule and Dieterlen ever arrived. The sequence of events, as imagined independently by Carl Sagan, Ian Roxborough, and myself, might run as follows. As I've tried to emphasize, Sirius B was important and widely disseminated news in the 1920s. Some Jesuit priest reads about it in *Le Monde,* and then goes to Mali long before Griaule and Dieterlen.

"Tell us your myths," say the Dogon.

"Do you see that star?" replies the priest; "it is actually two stars, and the invisible star is the heaviest thing there is."

The Dogon promptly incorporate this information into their culture. And when the two anthropologists are told the secrets of the Dogon, all they get is a cross-cultural translation. Now it seems to me that this is no doubt the most likely explanation for the knowledge possessed by the tribe. It counts in favor of this explanation that the Dogon say there is a third member of the Sirius system, a third star that orbits Sirius A in the same sense and with the same period as Sirius B. There were occasional reports of a third star in Sirius throughout the 1920s. It was the last time at which the separation between Sirius A and Sirius B was maximal: 11 arc-seconds. Accordingly, many astronomers were looking at Sirius. About half a dozen of them reported in the technical literature that they had seen a third star. Some astronomers even computed an orbit and a period for this so-called Sirius C, and claimed to be able to find its perturbations upon the visible Sirius A. None of this was ever confirmed, and yet, of course, our imagined Jesuit might have passed it on to the Dogon. (Fifty years have now passed since the 1920s, and Sirius A and Sirius B again are separated by 11 arc-seconds. About five years ago, I. W. Lindenblad, an astronomer at the US Naval Observatory, tried to find the third star. He couldn't. Moreover, he couldn't find any perturbations in Sirius A beyond those caused by Sirius B.)

I am told by anthropologists that cultural transfer of the sort I am supposing could not possibly have happened; there is no chance, they say, that missionaries, ethnographers, and the like could have smuggled modern notions into the core of a sacred tradition, which is what a cosmology is. I wonder about their certainty, though, for one can point out several examples in which cultures change wholesale. Consider the Cargo Cults of the South Sea Islands. After World War II, the islanders built large straw airplanes, hoping to lure the great birds from the sky—the birds which no longer were bringing them the products of "civilization." The straw airplanes are the central element of their religion. Consider, too, the cultural changes caused by a single man, D. Carleton Gajdusek, who received the 1976 Nobel Prize in medicine for his work on slow-acting viruses of the brain.

When Gajdusek visited various tribes in New Guinea in the course of his researches, he carried with him a microscope. Each tribe would tell him a different theory for the origin of disease. Then Gajdusek would show the tribe some bacteria, or perhaps an amoeba. Gajdusek wrote an article about his anthropological findings, and some Australian anthropologist decided that it was incorrect: all the tribes, the anthropologist thought, ought to have the same idea about disease. To test his conviction, he followed the trial of Gajdusek and interviewed the tribes that Gajdusek had studied. Each tribe *did* give a single theory—the theory that Gajdusek had taught them. Gajdusek himself tells the story that one night in New Guinea he and a Russian scientist sang for their dinner. They sang a Russian song, "Black Eyes." Some years later, Gajdusek chanced on a recording of the tribe's songs. Among them was one sounding suspiciously like "Black Eyes."

Suppose, despite all this, that cultural transfer is somehow ruled out—as it would be if it turned out that someone had visited the Dogon prior to about 1910 and had found their mythology to include Sirius B. Even then, the Dogon might have been told about its 50-year period—but not about its incredible density. It would be a great mystery how they could have known this latter fact. One possible explanation is that the Dogon traded heavily with the Egyptians. There are in fact many similarities in their cultures. It may then be that the Dogon got the information from the Egyptians at some time in the past. Of course, that would only push the essential quesion backward in time: for how, then, did the Egyptians know that Sirius was a binary system that includes a white dwarf? One obvious solution to the problem is discussed in a book, *The Sirius Mystery,* written by a man named Robert Temple, who spent eight years of his life working on the Dogon mystery. The theme of Temple's book is summed up by the following question, posed on the jacket of the book: "Was Earth visited by intelligent beings from a planet in the system of the star Sirius?" The case he makes for an affirmative answer to this question is even less credible than the arguments advanced for ancient extraterrestrial visitors by Erich von Däniken. For if visitors from Sirius had bothered to set down on earth, they should at least have had the courtesy to relay to the lucky greeting party the intelligence that Jupiter has at least 14 moons (not four), and that Uranus (as well as Saturn) has rings—half a dozen of them (a fact discovered only recently by earthly astronomers).

I am still convinced that cultural transfer, say from a Jesuit priest, is a likely explanation for the knowledge possessed by the Dogon. But if that explanation were disallowed, I would offer two final possibilities. The first of them concerns sheer probabilities. Suppose there are a thousand cultures in the world. Suppose further that anthropologists study all of those cultures. Each one of them has myths. Most of the myths conflict with scientific theory; a few do not. We are dealing here with the myth that is most nearly correct by our standards of truth.

This argument is not very satisfying, of course, when one asks how likely it is for the Dogon, *ex nihilo,* to have postulated a 50-year period, an elliptical orbit, and an immense density—all for an invisible star. I'd like to quote, however, from two accounts of a similarly unlikely achievement.

The first account appears in *Gulliver's Travels,* written by Jonathan Swift in the 1750s. Gulliver, in the section from which I quote, is visiting Laputa, a small island suspended in air by a giant lodestone. Among the population are great astronomers who meditate a great deal about the universe. "The lodestone," reports Gulliver, "is under the care of [these] astronomers, who from time to time give it such positions as the monarch directs. They spend the greatest part of their lives in observing the celestial bodies, which they do with the assistance of glasses far excelling ours in goodness. For although their largest telescopes do not exceed three feet they magnify much more than those of a hundred amongst us and at the same time show the stars with greater clearness. This advantage hath enabled them to extend their discoveries much further than our astronomers in Europe. They have made a catalog of ten thousand fixed stars whereas the largest numbers of ours do not contain above a third of that number. [This is an accurate remark about the state of European astronomy in 1750.] They have likewise discovered two lesser stars or satellites which revolve about Mars, whereof the centermost is distant from the center of the primary planet exactly three of the diameters and the outermost five. The former revolves in the space of ten hours and the latter twenty-one and a half, so that the squares of their periodical times are very near the same proportion with the cubes of their distance from the center of Mars. Which evidently shows them to be governed by the same law of gravitation that influences other heavenly bodies."

Now Mars does indeed have two moons: Phobos and Deimos. They were discovered only a century ago—a century after Swift wrote *Gulliver's Travels.* The actual periods of Phobos and Deimos are 7.5 hours and 30 hours. How did Jonathan Swift do so well? A clue can be found in an equally striking passage in Voltaire's *Micromegas:* "In one of those planets which revolve around the star named Sirius" writes Voltaire, "there was a young man of much wit with whom I had the honor to be acquainted during the last visit he made to our little anthill. He was called Micromegas, a name very well suited to all big men.

. . . When they left Jupiter they crossed a space about a hundred million leagues wide and passed along the coast of Mars, which as we know is five times smaller than our little globe. They saw two moons which serve the planet and which have escaped the attention of our astronomers."

The source of these remarkable guesses by Swift and Voltaire appears to be contained in a letter from Kepler to Galileo more than a century earlier: "I am so far from disbelieving," wrote Kepler, "in the existence of the four circumjovial planets that I long for a telescope to anticipate you, if possible, in discovering two round Mars (as the proportion seems to me to require), six or eight round Saturn, and perhaps one each round Mercury and Venus." The point is that Kepler evidently felt that the satellites (as well as the planets) should be arranged according to some geometric progression. One should not forget, by the way, if one wishes to become a scholar of coincidences and lucky guesses, that Voltaire's Micromegas hailed from an immense satellite of Sirius.

The last possible explanation of "The Sirius Mystery" is a broad and implausible one, lying in the realm of astrophysics. Its sole merit is that it accounts both for the ancient report of Sirius as red and for the Dogons' knowledge of Sirius B. Suppose that about 2,000 years ago Sirius B was indeed a red giant, so that in consequence Sirius A and Sirius B were roughly comparable in brightness. As they moved around each other, one would see a single point of light, but its color might change over a period of 50 years. If the red star dominated during your lifetime, you would call Sirius red. If you saw the colors changing you might say that there were two stars. In any case, you would learn of a 50-year cycle. And as the red disappeared, you might also decide that the star is growing old and shriveling up. By this argument, you might be led to believe that the second star, now invisible, is very dense. All of this may have been part of Egyptian cosmology, passed on to the Dogon, where it is preserved today.

Even with such a scenario, the redness of Sirius would create an incredible problem for theoretical astrophysicists, whose understanding of stellar evolution comes as much from computer calculations as from observations of the sky. It's my secret hope, nevertheless, that Sirius truly *was* a red star in historical times. I would much prefer to learn stellar evolution from the ancient myths of man than from the modern myths of the computer.

For further reading:

Ancient Astronomy in General: Four books are recommended: *Hamlet's Mill,* Giorgio de Santillana and Hertha von Dechend, Gambit Press, Boston, 1969; *The Dawn of Astronomy,* J. Norman Lockyer, MIT Press, Cambridge, 1973; *The Exact Sciences in Antiquity,* Otto Neugebauer, Harper Brothers, New York, 1962; and *Science Awakening II: The Birth of Astronomy,* B. L. van der Waerden, Oxford University Press, New York, 1974.

The Red Color of Sirius: This problem has been discussed many times over the last thousand years. Early investigations are summarized in Alexander von Humboldt's book, *Cosmos,* vol. III, 131-136, Harper Brothers, New York, 1851. T. J. J. See (whose life-story, entitled *Brief Biography and Popular Account of the Unparalleled Discoveries of T. J. J. See,* offers as its motto: "The Simple Truth—The Best Inspiration to the Youth of the Land") summarized many more arguments in the *Astr. Nach.,* **229,** 245 (1926). The astrophysical problem has been discussed by H. M. Johnson in *A.S.P. Leaflets,* 383 (1961), and by Zdenek Kopal in *Close Binary Systems,* Wiley & Sons, New York, 1959.

Dogon Astronomical Myths: M. Griaule and G. Dieterlen, "Un Systeme Soudanais de Sirius," *Journal de la Société des Africanistes,* **20,** I, 273 (1950), is the original article on the Dogon's knowledge of Sirius. More recent interest seems to have been kindled by two articles by W. H. McCrea in the *Journal of the British Astronomical Association* **84,** 63 (1973) and *Quarterly Journal of the Royal Astronomical Society* **13,** 506 (1972). Finally, a lively correspondence concerning the Dogon Sirius myths has appeared in the pages of the British magazine *The Observatory,* **95,** 52 (1975); **95,** 215 (1975); and **97,** 26 (1977).

OWEN GINGERICH

THE BASIC ASTRONOMY OF STONEHENGE

About twenty years ago, when I first saw Stonehenge, I was taken by surprise. Somehow, in mind's eye, the trilithons and stone circle had assumed truly monumental proportions. In contrast, the real stones were set on a disappointingly small scale. My confession is not intended to belittle the world's most famous megalithic site, but simply to report an honest reaction to one of the most romantic of all prehistoric ruins.

My initial response, that Stonehenge seemed unexpectedly small, is in fact relevant to the viewpoint that I should here like to defend: that Stonehenge is not so much an ancient megalithic observatory as the *monument to an earlier observatory.* By this I mean that any astronomical sighting lines at Stonehenge must have been well established centuries before they were fossilized into such a heavy, immobile configuration, and that the organization of the monumental stones is primarily dictated by the aesthetic symmetry along their principal axis and not by a secondary series of lunar sightlines, as some have proposed. At best, Stonehenge was a ritual center commemorating bygone discoveries, not a site where new knowledge of the heavens was actively sought. It was a stable monument to the eternal order and regularity in the sky, and as such its alignments still synchronize with the sun's rhythmic march through the seasons.

The statistics of Stonehenge are impressive enough. The stone ring, about 30 meters in diameter, originally comprised 30 megaliths capped by 30 lintels. Within this so-called sarsen circle stood the five huge trilithons in a horseshoe pattern; each was a pair of uprights with a lintel held in place by a mortise and tenon carved in the stone. The largest of the megaliths, one of the trilithon uprights, must have weighed 50 tons; it is the largest prehistoric hand-worked stone in Britain. By comparison, the stones of the sarsen ring are a mere 25 tons each. The sarsen stones—originally huge natural boulders but dressed by pounding with stone hammers—came from the Marlborough Downs, nearly 20 back-breaking miles north of Stonehenge.

Today nearly half of the sarsen ring has been quarried away, and only three of its stones remain untouched in their original positions. However, 16 are now in place and six have regained their lintels with an assist from the archaeologists. The trilithons have escaped the ravages of time somewhat more successfully. The two southeast groups never fell. When the largest group, on the central axis, tumbled down in ages past, one of the stones broke in two, but the surviving upright has been reerected in its original socket hole. The original orientations of the rings and the outlying, so-called Heel Stone, about 80 meters to the northeast, can be established reasonably well (figures 1 and 2).

Before we can understand the unique geometry of Stone-

Figure 1
Stonehenge from the east. The photograph was taken by the author in 1958; since then, some further reconstruction of the site has been done. Near the middle of the silhouetted image of the monument is a stone with a small nub on top. The stone is the remaining megalith of the central trilithon, and the nub is a tenon—a protrusion that once fitted into a corresponding hole in a lintel stone above.

Figure 2
Part of the sarsen circle at Stonehenge. Framed between the two visible sarsen stones is the Heel Stone, 80 meters from the center of the ring. The smaller stones in front of the sarsen stones are two of the so-called Blue Stones, not discussed in this article. Nothing is known about their function, except that they are placed along a circle within the sarsen circle, and that the Stonehengers themselves rearranged them. The stones in the foreground and at the left are fragments; many such stones are present at the site.

henge and the special orientations of other ancient monuments, we must grasp a basic idea about the cyclic motions of the sun, moon, and stars. I shall use Stonehenge as a specific example, but the general rules will apply whether the monument is at the latitude of Teotihuacan in Mexico, Karnak in Egypt, or Moose Mountain in Saskatchewan. As we shall see, however, there is a dependence on latitude that makes certain phenomena much more striking in northern realms such as the British Isles or the Great Plains of Canada.

Without any doubt, the relevant celestial motions were much more easily explained when I spoke at MIT in January 1977. There I had a live and captive audience, sufficiently enthralled with the romance of ancient astronomy to have braved a winter snow storm. It was far easier, too, to wave my arms in three dimensions and to spin a large armillary sphere, than to convey the same concepts on a flat printed page. As I contemplated the difficulties of this presentation on paper, I could all too easily imagine my readers laying aside their volumes and quietly sneaking off to their refrigerators.

And then inspiration struck. As a model of the celestial sphere, a cylindrical cola can is a fair approximation. The celestial equator, which splits the sky into its northern and southern hemispheres, nicely girdles the can. Cocked at a 23° angle to the equator is the ecliptic, the other great celestial circle that concerns us when considering the basic astronomy of Stonehenge.

To convince any skeptical readers that I am really serious about this analogy, I have provided a cutout wrap-around (see figure 3 and the appendix to this volume) that can easily be taped to a beer or soft drink can. (If you lack either of these, a Campbell's soup can has the same diameter.) To be thorough, I should have included a round paste-on for the top of the can with a dot in the center labeled "North Pole," but I decided that some things could be left to the imagination.

With the wrap-around taped to the cola can, the sun's daily motion around the celestial sphere can be imitated simply by rotating the can about its cylindrical axis. The sun's *yearly* motion among the stars, in the opposite direction, is shown by the succession of monthly images along the ecliptic. You can think of the sky as a giant celestial roulette wheel, with the heavens spinning eastward one revolution per day and the little ball, our sun, going around the other way, making one passage through the ecliptic each year.

The sun reaches its northernmost position near the end of June, but notice that it is almost equally far north in May or July. By our present calendar the sun reaches its highest point around June 21, the summer solstice, also known as midsummer day. It isn't midsummer by our seasonal conventions, but rather the *beginning* of summer. The arrival of the sun at this northernmost position is traditionally associated with some sort of a holiday; and Philip Morrison assures me that July 4 is, in fact, our national holiday for midsummer. When I objected, saying that everyone knows the Fourth of July marks the signing of the Declaration of Independence, Professor Morrison replied that "if it weren't so, there would have been some other holiday. There *has* to be a holiday at midsummer, and just at this moment in our national life, July 4 has captured the position."

As for the stars on the cola-can universe, I have omitted them, because the positions of the constellations today are quite different from those in the third millennium B.C., when Stonehenge was built. Nowadays, for example, when the sun is at its northernmost solstitial point it crosses the northwest corner of Gemini. But a few centuries before Christ, when the constellations were being named, Cancer occupied this place—hence the name "Tropic of Cancer." And when the anonymous men of prehistory laid out Stonehenge, the sun went near the star Regulus, in Leo, at the time of summer solstice. In another 9,000 years Regulus will be about as far south of the equator as it was north in 2000 B.C. The slow shift in the constellations is called precession. As a consequence of this shift, the zodiacal figures last matched their present positions 26,000 years ago, and during the next 26,000 years they will cycle all the way around the ecliptic and come back to their current coordinates. It follows that the rising direction of a bright star is a comparatively ephemeral orientation. If the pyramids had been built to line up with the rising of the star Sirius, they would now be quite askew with respect to that star.

On the other hand, although the stars gradually change their orientation, the 23° (almost 24°) tilt of the ecliptic stays pretty nearly constant, so the north-south excursion of the sun is nearly the same year after year, millennium after millennium. Stonehenge, built over 4,000 years ago to line up with the northernmost rising of the sun, still lines up with the northernmost rising of the sun.* In order to see *where* this northernmost point lies along the horizon, I have designed a horizontal plane with an elliptical hole in which to rotate the celestial cylinder. (See figure 3; you will have to cut out the ellipse, and another sheet is provided for this purpose in the appendix to this volume. I would recommend pasting the sheet on a thin piece of cardboard to give a greater planar stability.) The can must be tilted in the hole to fill the ellipse exactly; two black strips on the wrap-around will help maintain the proper slant while you turn the can.

The particular slope of the hole—and the corresponding tilt of the cylinder—depend, of course, on the geographic latitude one wishes to depict. For example, to represent an observer at the North Pole, the can would be straight up and down, and

*In fact, the tilt of the ecliptic *does* change slowly, and the reason the sun still lines up so well at Stonehenge is that the Heel Stone has tended to tip over in the correct direction.

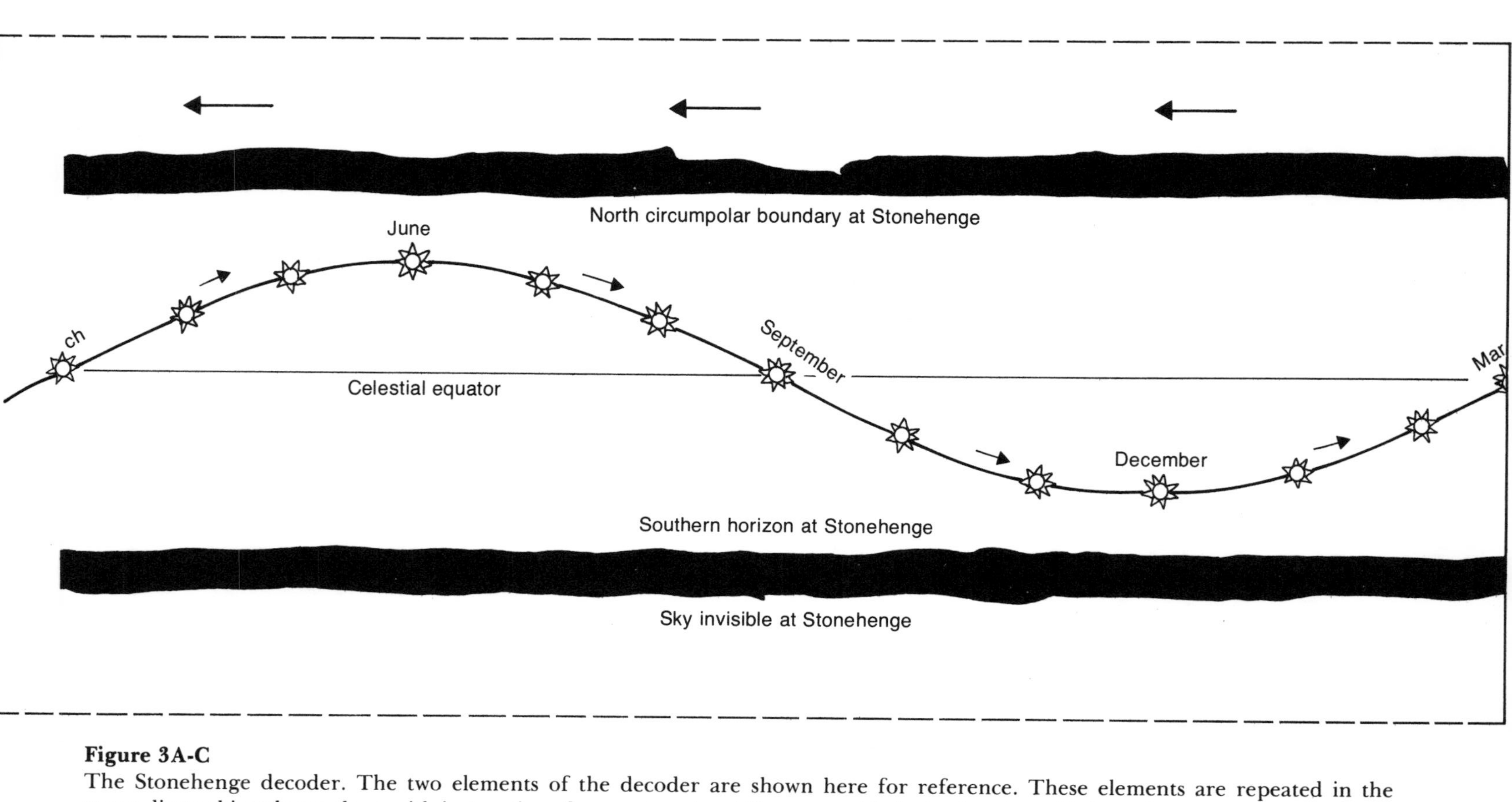

Figure 3A-C
The Stonehenge decoder. The two elements of the decoder are shown here for reference. These elements are repeated in the appendix to this volume along with instructions for construction. When assembled, the decoder will show the year-round motion of the sun over Stonehenge. Figure 3C (p. 124) is a sketch of the assembled decoder.

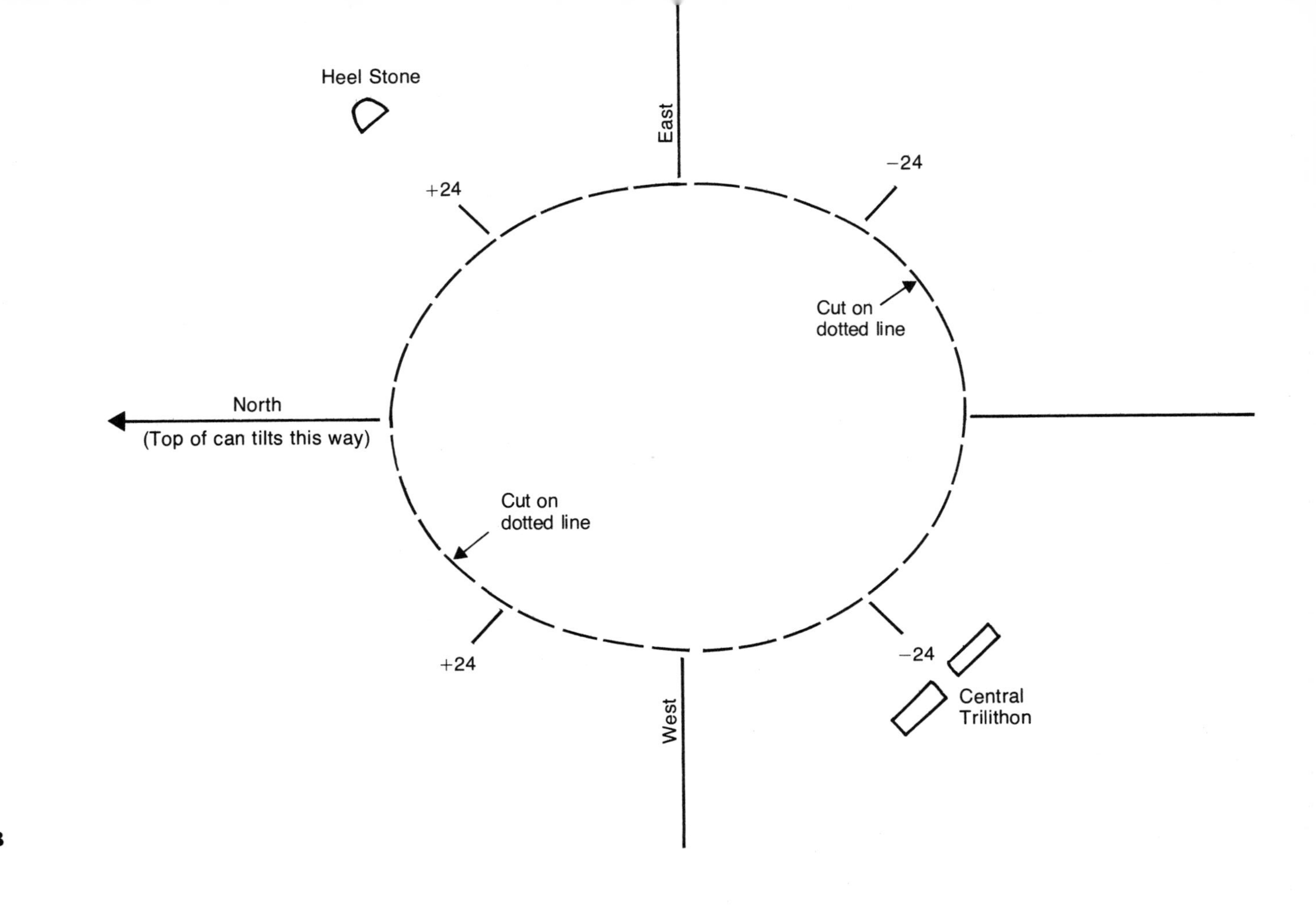
Heel Stone
East
+24
−24
Cut on
dotted line
North
(Top of can tilts this way)
Cut on
dotted line
+24
−24
West
Central
Trilithon

Figure 3B

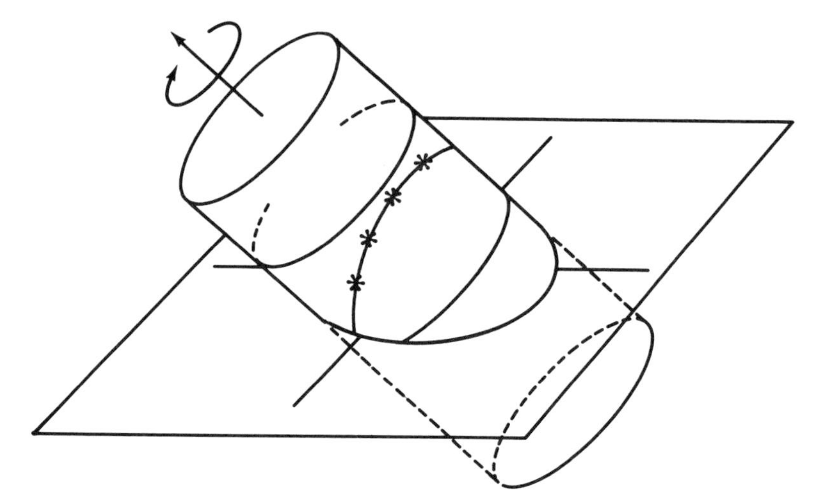

Figure 3C

the horizontal plane would have simply a circular hole cut into it. The sky north of the equator would always be visible, the part to the south never, so that clearly the sun would be in the visible hemisphere only during the spring and summer; the sun would circle the sky uninterruptedly throughout those six months. So much for monument building at the Pole.

For an observer at the equator the can would be on its side, and over the course of the year the sun would shuttle between 23° north of east and 23° south of east in its daily risings. At noon the sun would always be high in the sky, never more than about 23° from the zenith; and at noon-time twice a year it would be directly overhead. The annual excursion of the sun becomes larger and therefore increasingly interesting as an observer moves from the equator and tropics to more northern latitudes. On the model, the can has less and less tilt.

At an angle of 51.2° (the latitude of Stonehenge), the sun swings back and forth over an arc of nearly 90°. If you rotate the can in the cut-out ellipse (which is designed to represent the latitude of Stonehenge), you will discover how the rising and setting positions of the sun oscillate north and south along the horizon, and you will see that the swing is quite pronounced. Precisely how large is it? You can measure it on the model, or you can put it on your pocket calculator: $\sin D = \sin \delta / \cos \phi$, where δ, the sun's maximum angle above the equator (its maximum declination) is about 23.5° and the latitude ϕ is 51.2°. The maximum deviation D from the east-west line is just about 40°.

Enough for the basic mathematics. Let us next apply a thought-experiment to prehistory.

Ancient man, living on the island that is now called Britain, must have been aware of the rising sun's endless excursions back and forth along the horizon. Eventually someone must have marked by trial and error the direction of the northern rising of the summer sun. How fascinating, then, to watch day after day on some later year as the sun worked its way toward its northern limit. On each successive day the sun would take a smaller step than the day before, preventing any intuitive extrapolation. Would the sun reach the same limit? Or would it go beyond? What an exciting discovery to find that each year the sun reached *exactly* the same northernmost alignment! The weather might vary in a capricious manner, but the sun had a faithful regularity. Surely a discovery worth a monument!

The grand edifice at Stonehenge indeed points to the northernmost rising of the sun, for from the center of the stone ring, the sun can be seen to rise above the Heel Stone at the time of June solstice. In some of his more dogmatic moments, the distinguished archaeologist of Stonehenge, R. J. C. Atkinson, has denied this, and he was made to look rather silly several years ago when a CBS television crew filmed the sun rising majestically past the Heel Stone at the summer solstice. Still, the sun in truth rises to the left of the Heel Stone and moves to a position above the tip of that boulder. Hence, as Atkinson correctly claimed, the shadow of the Heel Stone cannot fall into the center of the ring at the solstice. In 2000 B.C. the sun would have risen a degree farther north than it does now, because the tilt of the ecliptic does vary slowly. Probably, though, the Heel Stone was more upright then; and in any event, a sightline of only 80 meters from Heel Stone to the center of the sarsen circle means that the observation is not very precise: the sunrise would appear almost the same for a week before and after solstice. Certainly for ritual purposes the principal axis of the monument points to the rising sun at the summer solstice. By symmetry the opposite direction points southwest, to the southernmost setting of the sun at the winter solstice.

The real question about Stonehenge is, does it point to anything else? More specifically, would it have required a Stone

Age Einstein to think of asking whether the *moon* also had a northernmost rising point? I am not sure. What would have been needed for investigation of the moon was a sufficient time free from the exigencies of food gathering. But a special sort of curiosity would have been necessary as well. Each month, in one phase or another, the moon will rise once in the vicinity of the Heel Stone, yet in general this alignment will not be very exact. It thus could be ignored as one of the vagaries of that fickle object. Furthermore, not all of these risings would be visible; the solar brilliance at midsummer would overpower a thin crescent rising soon after the sun, for instance. Finally, because the moon moves about 30° a day, it could skip past the very northernmost position on any given cycle.

The reason that the moon's northernmost rising is only in the vicinity of the Heel Stone, and not dead on, is that the moon's path is askew to the sun's. (If it were not, there would necessarily be a lunar and solar eclipse every month.) The moon's path in fact cuts the ecliptic at two opposite points in the sky, and wanders off by 5° north of the ecliptic on one side of the sky, then 5° south on the other. Moreover, the place where the maximum wandering occurs is slowly changing. Hence the moon's path can go 5° north of the 23.5° northernmost point of the ecliptic, but nine years later it will go 5° south of this northernmost point. In the first situation the moon would move in the sky between 28.5° north and 28.5° south, and in the second between only 18.5° north and 18.5° south. With the handy formula we used earlier in this article, we get an extreme swing along the horizon of 50° north of east, well past the Heel Stone. But nine years later, we find that the moon at most would rise only 30° north of east. In other words, our Stone-Age astronomer would not only have to have the insight to ask if the moon, like the sun, had a northernmost rising, but he would also have needed plenty of time to be sure—18 years to complete the observation of just one complete cycle.

Would it be easy to think of trying this? I don't know, *but we can be rather sure that it was done.* Excavations at Stonehenge have revealed a series of postholes at angles northward of the Heel Stone in just the positions we would expect if someone was trying to establish the moon's northernmost limit by trial and error (figure 4).

In its earliest stages the Stonehenge site was surrounded by an embankment approximately two meters high and about 100 meters in diameter. This earthwork ring had one entrance, in the general direction of the Heel Stone, but not symmetrically aligned with it. The opening was just wide enough to accommodate the postholes lined up with the northernmost excursion of the moon. Because similar grids are found at other neolithic sites, there is good reason to suppose that these represent lunar alignments.

Gerald Hawkins, who more than anyone else has drawn attention to the astronomical nature of Stonehenge, claims that solar and lunar alignments can also be associated with the four trilithons flanking the main axis of the monument. These are shown in figure 5, where the megaliths are represented as idealized rectangles. The idealized representation is in itself instructive, because we notice at once that not all the sightlines would be possible on account of the awkward slanting views required. To be sure, the actual megalithic boulders are more rough-hewn, so the sightlines are in fact feasible. Yet it gives pause to me, at least, to suppose that a grand edifice of this sort was specially designed so that the sightlines would depend on the imperfections or asymmetrical sizes of the rocks. I would also worry if the southern moonrise extremes and the northern moonset extremes were megalithically marked, and not the other pair of events (northern moonrise and southern moonset).

Up till now I have virtually ignored one important historical aspect of the Stonehenge site: the monument we now admire is Stonehenge III, built around 2100 B.C.; the sarsen ring and trilithons stand in the center of a much larger and older site whose development occurred hundreds of years earlier. And it is the early Stonehenge that has a greater claim as a research observatory, if for no other reason than that the sightlines are considerably longer and more flexible than those proposed by Hawkins for the sarsen ring and trilithons. The early Stonehenge included the previously mentioned embankment, as well as a ring of 56 chalk-filled holes, now named after the antiquarian John Aubrey. At some point in the development of the site there was added onto the ring of Aubrey Holes two mounds and two additional marker stones, known as the station stones (see figure 4). These four features form the corners of a rectangle centered on the sarsen ring, although the sarsen ring had not yet been built at that time. The short sides of the rectangle are parallel to the direction to the Heel Stone; hence

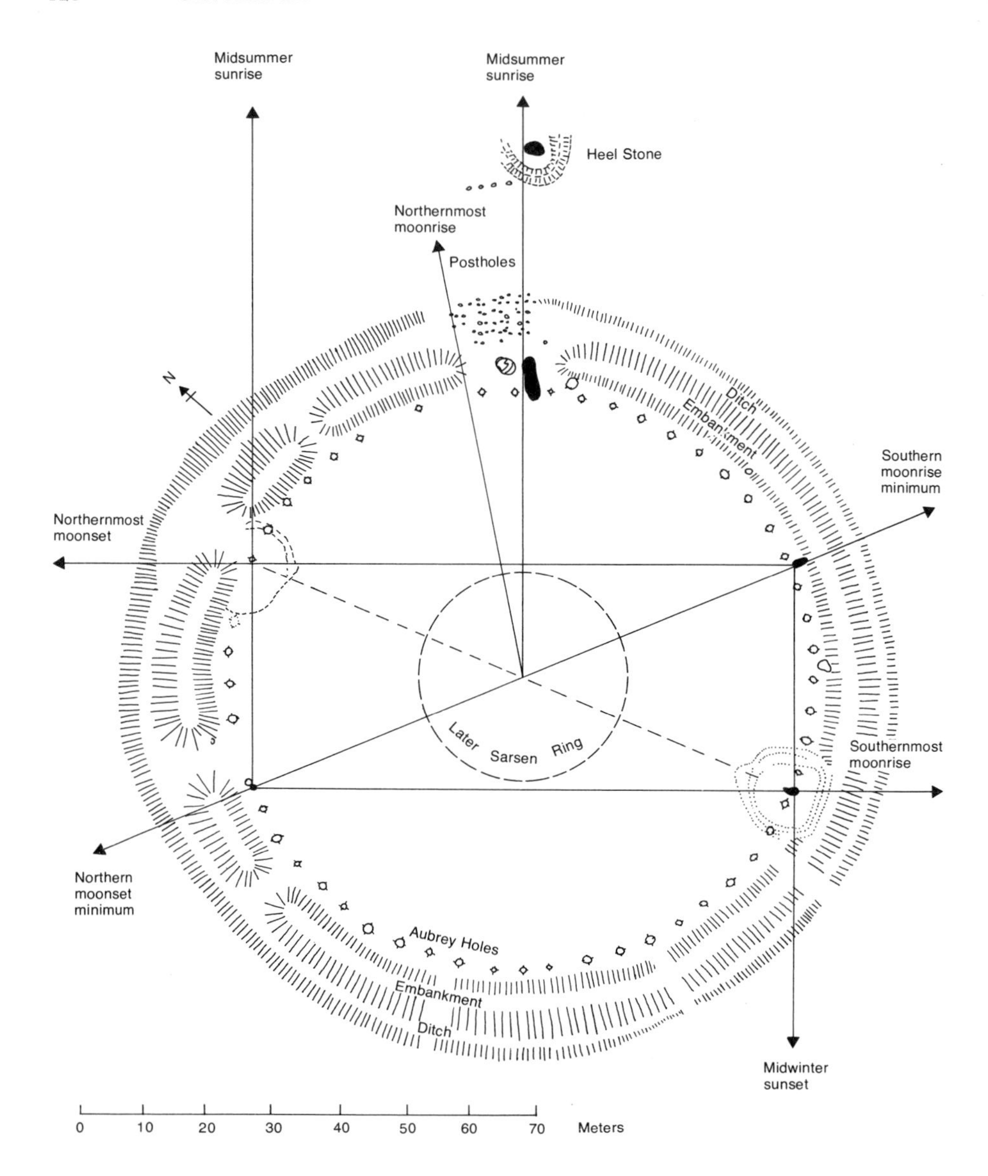

Midsummer sunrise
Midsummer sunrise
Heel Stone
Northernmost moonrise
Postholes
Ditch
Embankment
Southern moonrise minimum
Northernmost moonset
Later Sarsen Ring
Southernmost moonrise
Northern moonset minimum
Aubrey Holes
Embankment
Ditch
Midwinter sunset
N
0 10 20 30 40 50 60 70 Meters

Figure 4
The early Stonehenge—or rather, the early Stonehenges, for the illustration shows several stages of construction at the site. The first of these—"Stonehenge I"—is an earthwork ring about 100 meters in diameter and two meters high. Its completeness was broken (as of about 2850 B.C.) by a single gap directed in the approximate direction of an outlying marker called the Heel Stone; and in this gap, excavation has uncovered a grid of postholes: the remains, it seems, of an effort to mark the northernmost excursion of the moon. Note that the Heel Stone lies slightly away from a line drawn from the center of the earthwork ring to the horizon point marking the midsummer (solstitial) sunrise; in 2850 B.C. the Heel Stone was presumably more erect, and thus the alignment was more nearly perfect. Stonehenge I also included a circle of chalk-filled holes now named after John Aubrey. At some later time, Stonehenge II was added. It includes low earthwork constructions that center on the ring of Aubrey holes, and also the so-called station stones. As shown in the illustration, these additions to the site mark out the corners of a rectangle whose sides and diagonal align with various risings and settings of the sun and moon. In about 2100 B.C., Stonehenge III was constructed at the center of the site (shown by the circle of dashes). Stonehenge III is the megalithic structure that draws our attention to the site today.

they point to the extreme excursion of the sun. The long sides point to the extreme excursion of the moon. (Only at the latitude of Stonehenge are these extremes joined by a right angle; if this is significant, then these relationships were not discovered here, but the observatory was erected at this specific latitude to take advantage of a previous discovery.)

One diagonal points to the moonrise and moonset when the moon has the *least* excursion north and south—scarcely an obvious phenomenon, and one that would take considerable sophistication to recognize and mark. However, A. R. Thatcher has recently pointed out that the diagonal is even more accurately aligned to the May Day sunset, the time halfway between the spring solstice and midsummer. Such lines are found in a number of stone circles, and would correspond to the later Celtic Beltane festival.

The longer lunar alignments (if so they be) in the early Stonehenge do have sightlines an order of magnitude greater than those from the trilithons to the sarsen ring. Even so, they are short compared to the megalithic sightlines of 18 and 27 miles that Alexander Thom has surveyed at Ballochroy and Kintraw in Scotland. Since Stonehenge seems to be an older site, perhaps

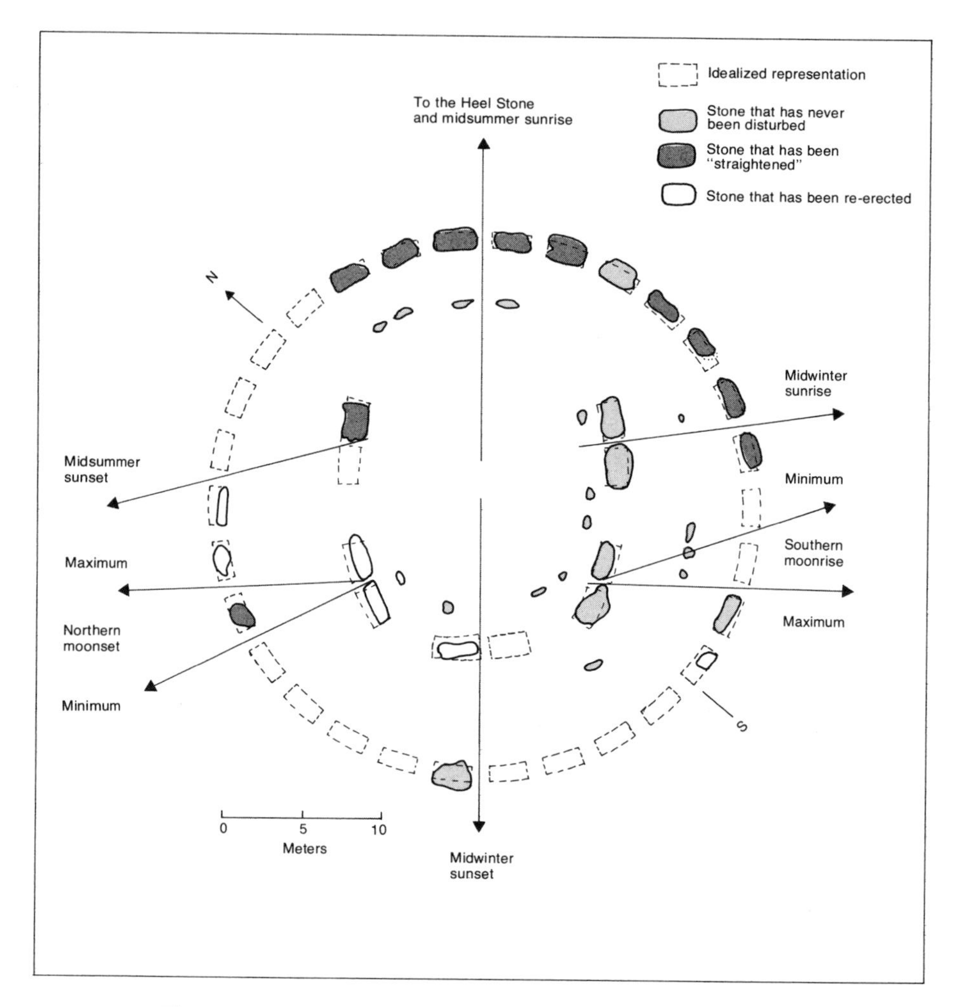

Figure 5
The later Stonehenge—Stonehenge III. Dotted lines show what one time stood at the site: a horseshoe of five trilithons within a ring of 30 sarsen stones. Eighty meters from the center of the ring is an outlying marker, the Heel Stone; in combination with the central trilithon, it marks the central axis of the monument. The illustration shows several putative alignments of the monument to various risings and settings of the sun and moon. The two solstitial alignments associated with the central axis are far less problematical than the alignments associated with the off-center trilithons. The illustration fails to show the various fallen stones and fragments. Yet no rock of any size was originally present: the soil beneath the grassy surface is exclusively chalk.

the astronomy spread from there to the more accurate observatories elsewhere in Britain.

Indeed, Stonehenge cannot be considered in isolation, for the patient surveying of Alexander Thom and his associates has attempted to establish the existence of lunar sightlines in conjunction with many of the more elaborate stone circles in the British Isle, a fact about megalithic society now almost commonly accepted by the archaeologists. Yet in Stonehenge III we find no elaborated observing site with longer baselines and more precise markers than the earlier Stonehenge constructions. Instead there is a monumental commemoration in stone of something long since discovered and perhaps already on its way to being forgotten. There is a striking parallel to this in eighteenth-century India where the ruler Jai Singh created five impressive stone observatories, all completely anachronistic (considering that the telescope had been introduced into Asia decades earlier), yet comfortable to the monarch's aspirations. Indeed, Jai Singh's instruments remain impressive till this day.

The combination of lunar and solar sightlines embodied in a great ritual center at Stonehenge suggests a well-organized primitive cult, possibly with the sun and moon as male and female in some grand fertility rites. Such suggestions are mere speculation, for here the stones are even more silent.

Nevertheless, the idea is no more fragile than the proposition that Stonehenge was a Stone Age eclipse calculator. There is a lunar eclipse cycle of 56, albeit not a very good one, so the 56 Aubrey Holes could conceivably have been involved with eclipses. I beg the reader's pardon for not trying here to explain either the cyclic behavior of eclipses or the interesting suggestions by Hawkins and by Fred Hoyle for using the Aubrey Holes to calculate either the eclipse "danger seasons" or the eclipses themselves; the references at the end will give a full account.

Suffice it to say that I remain skeptical. There are certain aspects that stagger the imagination. To get the idea that you could predict eclipses in some cyclical fashion, you would have to have some long record of observations and some kind of motivation for recording them in the first place. Such a record would presumably have to be oral. Today we cannot begin to conceive of the significance of oral records. We have too much cluttery detail to remember, and we are not very good at memorizing things. I am sure that memorization must have played a much more significant role for ancient people than it does for us, because we are so dependent on the written record. Even so, such a route to the prediction of eclipses seems incredible to me.

On the other hand, for a people so concerned with capturing the northernmost position of the sun and moon, the conception of the moon's nodes (the points at which its path crosses the path of the sun and where consequently eclipses can take place) may not be terribly far behind. In other words, it may well have been possible for the Stonehengers to have correlated eclipses with the celestial geometry of the solar and lunar paths rather than with cycles of eclipses derived from a communal memory of events long past. To me, this seems to be a fabulous jump for neolithic man to have made, but there is nothing to have prevented a Stone Age genius from finding the correlations simply with sticks and stones. So perhaps one of these days we will have to revise our notions concerning the sophistication of megalithic astronomy in the third millennium B.C. Until then, in the words of Aubrey Burl, "This ravaged colossus rests like a cage of sand-scoured ribs on the shores of eternity, its flesh forever lost."

For further reading:

Stonehenge Decoded, Gerald S. Hawkins, Doubleday, New York, 1965.

Megaliths, Myths and Men: An Introduction to Astro-Archaeology, Peter Lancaster Brown, Taplinger, New York, 1976.

Science and Society in Prehistoric Britain, Euan W. MacKie, St. Martin's Press, New York, 1977.

"Stonehenge," Alexander Thom, A. S. Thom, and A. S. Thom, *Journal for the History of Astronomy,* **5,** 71–90 (1974).

The Stone Circles of the British Isles, Aubrey Burl, Yale University Press, New Haven, 1976.

On Stonehenge, Fred Hoyle, Freeman, San Francisco, 1977.

JEROME Y. LETTVIN

THE GORGON'S EYE

We seldom look at the sky on a starry night. Indeed, some of us will never see it at all, as we pursue our rational goals in the modern world's haze. But in preindustrial times the sky at night was an important matter. There is a lovely passage in *Don Quixote* in which Sancho Panza tells how a shepherd would know what time it was, and thus when dawn was approaching, by observing the position of a particular constellation. If you dip into the works of antiquity all the way back to the dimmest past of any human culture, you will find similar references, similar appreciation of the skies among people at large. It was not a knowledge confined to specialists.

How did the ancients talk about the sky? How did they identify the things that are in it? How did they keep track of those things night after night, year after year?

People who remember orally have no need to remember on paper, so one seldom finds those who are equally skilled at both, because each talent takes a great deal of time. Moreover, one seldom finds written accounts of what it was like to remember in the old-fashioned way prior to the time of writing. One finds only stories that are told and retold. Plato tells of a time when the god Thoth brought the art of writing to Rameses. "What will happen to our memory," Rameses sadly asks, "now that we can keep it on paper?" Thoth gives no answer.

Probably the only written work available to us about the ancient arts of memory is the *Rhetorica ad Arrhenium,* improperly attributed to Cicero. Here we find prescriptions for remembering things. We are instructed, for example, to imagine a walk through a temple, quiet and alone, during which we keep several feet from its wall. The argument that we wish to remember becomes associated with a sequence of things we see on the walk within the temple; and indeed the English word "topic" derives from *topos,* or place—a topic is the place in the temple to which we have attached an idea in the course of our walk. In effect, the temple comes to embody a kind of cross-mapping, in which something that we wished to remember superimposes itself upon something that we already knew very well. The temple is thus a stable and well-known ground on which the new set of facts is arrayed: the temple serves the purpose of a theory, to the extent that you can derive from a theory a great many facts, that, though seemingly unconnected, can nevertheless be related by the mapping you have made.

We have, therefore, an art of memory in which a well-known universe can serve as a ground upon which other universes can be mapped. But when you do such a thing, the world is no longer held together by what you might call conventional reason or logic. This leads to a curious kind of frustration when you seek to investigate the past. Consider, for example, that if you examine the Egyptian temples of Ra, or the temple supposedly devoted to the star Sirius, you will find that they are oriented upon very specific points on the horizon: the point of solstitial rising in the case of the sun, and the point of heliacal rising in the case of Sirius. (Other articles in this volume explain these matters.) Now given these temples, all oriented in special ways and all dedicated explicitly to the god of the sun or the star, one cannot easily argue that the alignments were accidental. And once you admit that the Egyptians made astronomical devices, you must also admit that the Egyptians had a prior knowledge of astronomy, even though that knowledge may not be translatable into the explicit terms that we ourselves employ. The problem is that the scholars who have studied the various temples find chiseled or painted on their walls what appear to be fairly unsophisticated stories: Ra did this, Ra did that, Ra returned. In short, the Egyptians inscribed picaresque stories on the cases of first-class scientific instruments, and this seems wrong-headed. It is as if one had unpacked an oscilloscope, looked in the box for an instruction manual, and found instead a book by Dr. Seuss.

Still, let us assume that the ancients spoke in parable, namely that what was said was put in rhetorical terms because of not having a language with the modern exactness—not having, that is, the terms in which to express even simple trigonometric relations, such as the position of the sun along the horizon on the day of summer solstice. And let me invent a fantastic astronomy so as not to scandalize the astronomers. Suppose I were to say that Floog suckled Goom. You would immediately have a feeling that the one preceded the other in time: that Goom was born from Floog, and therefore came later. This is a nice way to communicate the notion of a close time precedence, say, of a star appearing with respect to the sun at a place like the horizon. For you no longer have to remember an abstract relation about lights in the sky. You remember instead a concrete relation which is rich enough to include among its interpretations the particular one that you mean to emphasize.

I say, accordingly, that the myths of the ancients embody the relations among the stars spoken of by ancient astronomers. I say also that the myths may mean many other things besides. Myths, after all, are synthetic propositions, made in this world and subject to all the ambiguities and interpretations of any comment that isn't couched antiseptically in the precise language of modern science. But that only adds to the richness and, if you wish, the memorability of the ideas. For myths are more memorable the more things one can map on them. And there is for me a great poetic quality in a language whereby the relations of animals to each other, people to each other, the heavens to the earth, the gods to humankind, can all be worded in about the same way, until finally, by a single set of sentences, I can remember all of the universes as if they were maps of one another.

Before one can invent a myth about the skies, however, one must invent a myth about the earth. For the ancients had no way to speak of relations among mere points of light. Therefore, let us begin with something terrestrial, upon which the stars will later be mapped. Since the Greek myth of Perseus is considered to be rich in astronomical associations (about which more later), I will attribute to it a terrestrial significance.

Here is the myth itself. Perseus is born under strange circumstances: Acrisius, King of Argos, has been told by an oracle that he will be killed by his grandson. He has only a gorgeous daughter named Danae. He therefore locks her up in a brass prison so that she cannot be eyed by any man. But Acrisius fails to reckon with the lust of the gods, for Zeus sees through the prison wall, enters Danae's cell as a shower of golden rain, and seduces her. She becomes pregnant and bears a son: Perseus. Acrisius finds out about the birth and puts daughter and grandson to sea in a boat without oars. If the gods want to save Perseus (he argues), then *they* will do so; if not, *they'll* kill him.

Acrisius has nothing to do with it—a marvelously modern position.

The boat is blown ashore on the island of Seriphus, where Perseus is put up in the temple of Pallas Athena, to be educated and raised by the priests. As Perseus grows to teen-age, the local king, Polydectes by name, conceives a passion for Danae. But he makes no attempt at seduction, fearing reprisal from the youth. Perseus is in his teens when the king throws a party. Everyone who is invited is required to bring a gift horse. Perseus, who has no patrimony with which to get a horse, comes to the party nonetheless, and promises instead to give Polydectes the atom bomb of the time, the head of Medusa, which turns to stone anybody who happens to look at it. The king is pleased: either he'll get the ultimate weapon or he'll be rid of Perseus and have his will of Danae. One way or the other, he benefits greatly.

Perseus now sets out after Medusa. But first he goes by night to the temple of Pallas, where he is offered several gifts from the gods. Hades gives him a helmet of invisibility: it renders the wearer unseen. Hermes gives him the famous winged sandals and also a sword made of diamonds. Pallas gives him her shield, polished to mirror brightness. Perseus also receives a magic wallet, or kibisis, which can completely engulf whatever is put in it, however large, and yet not itself increase in size.

Provided with these devices, Perseus takes off in search of the three Gorgon sisters, of whom one is the Medusa. To find them, however, he must first inquire of the Phorcyades, three sisters of the Gorgons, who know where the Gorgons hide. The Phorcyades are like the landladies of Beacon Street: they have only one eye and one tooth that they pass back and forth among themselves. Invisible by virtue of his helmet, Perseus swoops down, snatches the eye and tooth, and promises to give them back only if the sisters will tell him where the Gorgons lie. This they do. Perseus returns the parts and takes off again. From a vast height, he sees the Gorgons, sunning themselves and sleeping at the seashore in Kisthene, the land of rock-roses. Of the three, Medusa alone is mortal. Gazing only at her reflection in the mirror shield, Perseus comes down backward, lops off her head, and thrusts it into the wallet. The remaining sisters wake up and cry. (Inspired by these sounds, Pallas will invent the flute.) From the neck of Medusa's body issues a storm cloud whence come her two children: Chrysaor, the Golden Sword;

and Pegasus, the Winged Horse. Perseus escapes by means of his magic helmet and sandals. He goes through a variety of adventures, using the head to turn kings into rock piles, rescuing and marrying Andromeda, and finally killing off his grandfather as was predicted. It's a beautiful story.

But what I wish to examine most closely are the Gorgons and Phorcyades. Who are all these sisters? Monsters from the id, from within the human unconscious, as if the ancient poets were H. P. Lovecraft or Carl Jung writing in a previous incarnation? (I have long felt that H. P. Lovecraft is Jung's nom de plume in English: both are devoted to calling up vague, subliminal feelings about things that are never described or expressed.) There is a remarkable amount of structure about the pair of awful triplets. For one thing, the Phorcyades, as well as the Gorgons, and indeed most of the other monsters that inhabit the Mediterranean, are spawned by Phorcys out of Ceto, Phorcys being the ancient god of the sea and Ceto his sister. Of the Phorcyades, nothing too much is said; they are the gray ladies. But about the Gorgons the poets are explicit. Now the poets were emphatically not advertising copywriters; they were people to whom words apparently meant something. For despite the fact that there were administrators at the time, language had not yet been overly corrupted. The Gorgons, say the poets, can turn from black to white and back again in an instant. They have living snakes for hair. Their gaze turns living creatures to stone. And they have "beautiful cheeks"—a curious phrase for beings so terrible. One further attribute: on one of the Greek Islands, there is a temple to the goddess Gorgo, a temple with the same form as that which the Parthenon was later given. And on the pediment of this temple is a tremendous Gorgon. Its tongue is sticking out. So also does it stick out for the Gorgons on medallions, vases, reliefs. The modern Greek scholars will assure you that this was a sign of fear or terror among the Greeks. But that is a strange sort of assertion; why should a tongue sticking out appear only on Gorgons and no other creatures?

Does any of this make sense other than as a tale to be marketed at an occultist bookstore? Let us begin with the gifts that Perseus received from the gods and move from them to the properties of the Gorgons themselves. Squid are common in the Aegean Sea. Suppose you take a squid and cut off its head. Its shape is now that of a sandal with wings at the toes—the flying sandal of

Hermes. And indeed, as Aristotle reported, the squid are known for flying: if a school of squid is attacked by a large fish or otherwise frightened, all of them shuttle in formation forward a short distance, and then, with a sudden and powerful contraction, force a strong jet of water out of their funnels to shoot into the air, still in formation. The entire school can travel thirty feet at a jump.

Consider now the helmet of invisibility—not of transparency, for the Greeks had a perfectly good word for that. A famous statue of Athena includes a helmet in the shape of a (headless) sepia—a cuttlefish. They, too, are common in the Aegean Sea. And what is interesting about the sepia is that it vanishes against every one of its surroundings; it is the most difficult animal to see under water, because it can change its color—and the pattern of its colorations—to suit any surrounding (figure 1).

Then we have the wallet that never grows larger, no matter how large its contents. Anyone who has ever watched an octopus feed will know of a resemblance here. I myself have given an octopus 35 crabs—an aggregate much greater in size than the octopus itself. Yet every one of them vanishes into the depths of the creature, and the octopus never grows larger. That is because the octopus first hides them all in the spaces under the interbrachial web between its outspread arms. This space is much greater in volume than the mantle (the body) of the octopus. The crabs lie immobile, for the saliva of the octopus has paralyzed them. Even then, however, the octopus does not precisely eat its prey. The saliva liquefies their flesh, and the octopus then sips them, as we sip coffee. The octopus, like the other creatures I have named, is common in the Aegean Sea.

I offer all this as prelude. The next step is to account for the poets' description of the Gorgons. Accordingly, I ask: what animals turn white to black and black to white in an instant? All the cephalopods I have just described: the squid, the sepia, and the octopus. What animals have snakes for hair? Consider the octopus with arms spread out above its staring, bulging eyes. And consider that if one of the arms is severed from the body, it will wiggle away on its own and stay alive and irritable for quite a while. What animals have beautiful cheeks? Put your hand under water, and you will find that the octopus blows water at you. Whatever blows can, with reason, be called a set of cheeks, especially since in the octopus these cheeks lie below the eyes, as the snaky locks lie above. And the cheeks of the

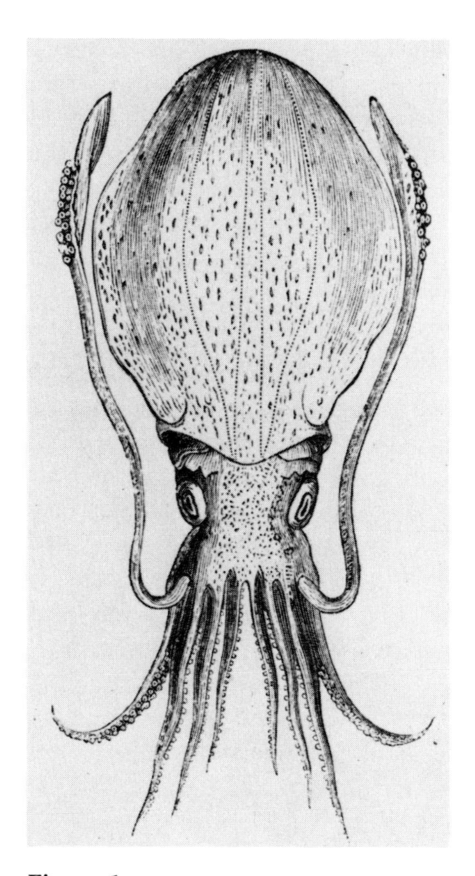

Figure 1
The common European cuttle-fish or sepia, *Sepia officinalis.* The creature, like its fellow cephalopods the octopus and the squid, boasts chromatophore organs over the entire expanse of its skin. The organs include pigmented discs, white, red, yellow, or black in color; and under the control of the nervous system, these colored regions can be revealed or hidden. The result is that the creature can abruptly change color, taking on a uniform, a mottled, or even a striped appearance so as to vanish against its background. The shape of the sepia, when the head and tentacles have been excised, is much like that of a helmet often depicted in Greek statuary.

octopus—which we less vividly call the mantle—are beautiful because they are colored and can change their colors. What about a Gorgon's tongue sticking out? Protruding from beneath the octopus is its so-called siphon, which has much the same mobility as the human tongue and can be pointed in any direction. It is always hanging out, and this is not a sign of fear; it is the hallmark of the cephalopods (figures 2 and 3). What about the exchange of eye and tooth, which Aeschylus attributes to the Gorgons as well as to the Phorcyades? The eyes of an octopus are backed up by a large blood sac or sinus, and when

Figure 2
The head of a Gorgon, a creature whose gaze turns men to stone. The head is ringed by writhing serpents, and out from the grinning mouth protrudes the awful creature's tongue. A tongue protruding in that curious fashion is typical of representations of Gorgons from antiquity. The drawing was made by Ruth McCambridge from an original fashioned in painted terracotta by Etruscans of the late sixth or early fifth century B.C.

two male octopuses meet side by side, the eye of the first protrudes as the sinus inflates, so that the eye seems to grow to four times its normal size. The eye of the second responds in kind, and finally the weaker one retracts its eye and goes away (figure 4). As for the exchange of a tooth, a fight between two octopuses will suggest an analogy. For first one octopus protrudes its beak and then the other does the same, as they struggle in arm to arm combat that looks like the Laocoön. If a tooth is a structure with which to bite, then here is the exchange of a tooth.

Only one attribute of the Gorgon Medusa remains—the most

Figure 3
The common octopus (*Octopus vulgaris*), which lives in warm waters around the world but is especially common in the Mediterranean Sea. Eight arms radiate outward from the creature's head. They are joined at their bases by a filmy web. On the head itself, two eyes can easily be seen. Just to the side of those eyes, the creature's siphon appears; it is a tube with the maneuverability of a tongue. The large, pouch-like structure beneath the eyes is the so-called mantle. The octopus thus far described is a female; the photograph also shows a male, lying at the bottom of the tank with one of its arms—the so-called copulating arm—in the process of fertilizing the female. The male's siphon is quite plainly visible. This photograph and that of figure 4 were provided by Professor Jerome Wodinsky of Brandeis University.

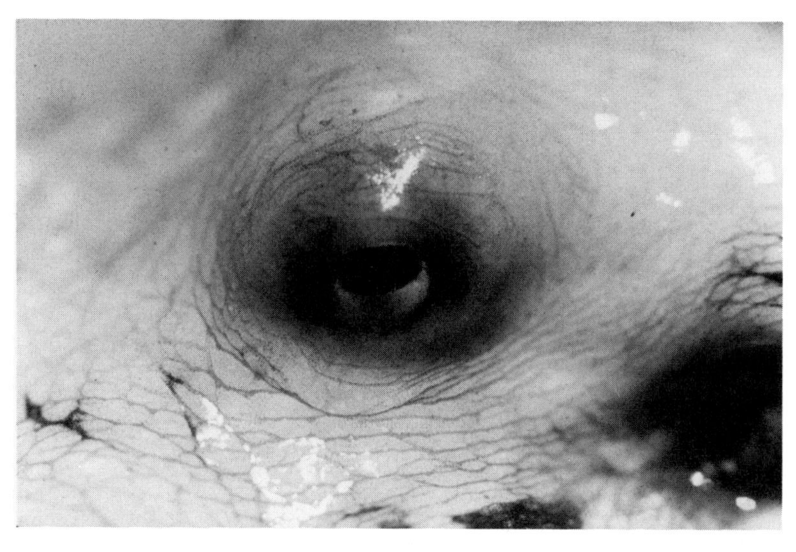

Figure 4
The eye of the common octopus, *Octopus vulgaris.* It is an organ much like the eye of a man, for both structures have a cornea, an iris, a fluid-filled chamber, a lens set in a ring muscle, and a tough fibrous sheath surrounding the eyeball. Moreover, both creatures have eyelids. Such similarities are most astonishing, considering that nature seems to have engendered the eyes of men and octopuses by entirely distinct lines of evolution.

well-known and horrific attribute of all. What animal has a gaze that turns other creatures to stone? Drop a living crab into a tank occupied by an octopus. The crab begins to scuttle about, the octopus opens its eye—and the crab stops in mid-flight! For if the crab moves even the slightest fraction of an inch the octopus will grab it. Its only way of hiding is to be motionless—and nature has arranged for it not to move.

Surely the Greek fishermen knew of the lives and habits of the squid, the sepia, the octopus. Here then is a story in which no fact is left dangling. Here are the daughters of Phorcys the seagod. Here are the three sisters; Aristotle called them all molluscs, and he was right: the cephalopods are the sapient cousins of the clams. Squid, sepia, and octopus are the only celphalopods on the shore of the Aegean Sea. They are memorable because when you look at them, you cannot help but remember their story. And when you remember them, you will find that you are walking quietly and alone in a frame of mind like a true temple. Along the wall of properties is a series of topics, a set of events or attributes which you can now ascribe to objects in the sky. You are edified, for no set of relations in the sky alone was sufficient to determine a theory, to build a thought temple, with the only language available to you.

Let us suppose that the Medusa, whatever meanings she may later have assumed, just possibly began with the description of an octopus. Does this help us at all in dealing with the astronomical counterpart of the myth—the constellation named Perseus? To be candid, I don't at first see that it does. Indeed, the more I regard the starry image, the stranger it becomes in this sense: there is no geometry that seems to hold between the stars of the constellation, and nothing about the position they enjoy in the heavens, so that the name Perseus should necessarily have been conferred upon them. Is the whole of the myth so abstract a mnemonic that any set of stars might have done as well?

I am going to suggest an answer that will seem brash to historians of astronomy, because the proofs lie in the absence of a piece of evidence—as with one of Chesterton's detective stories. But first consider the astronomy.

A little below the midpoint between the sword brandished by Perseus and the heel of his right foot in figure 5 lies a quartet of stars, the Gorgoneae, or the Gorgon's Head (*Caput Medusae*) in Ptolemy's catalogue. Even Perseus himself is not so fixidly named as is this Head. For all the sundry versions of the constellation, from the Cacodaemon of the astrologers to the al Ghul (The Bearer of the Demon's Head) referred to by the Persian Hamil Ra—all of these reflect a malignity at the constellation's center.

The brightest star in the Caput Medusae is called al Ghul. It appears as the right eye of the Medusa in the figure, and concentrates the malignity of the Demon's head to a single light whose usual magnitude is 2.3. Algol is also called Gorgonea Prima, but it is known to modern astronomers as β Persei, the second most luminous of all the stars in Perseus. Sometimes it looks red.

The next brightest star of the Medusa group is ρ Persei, or Gorgonea Tertia, whose maximum amplitude is 3.4. It is orange in color. The two dimmest stars, π Persei, or Gorgonea Secunda, of magnitude 4.5, and ω Persei, or Gorgonea Quarta, of magnitude 5, complete this segment of the constellation.

Now Algol, which is called Tseih She, the Piled-up Corpses,

Figure 5
Perseus holds the severed head of the Gorgon Medusa in a representation of the constellation Perseus taken from the *Firmamentum Sobrescianum,* a sky atlas published by Jan Hevelius in 1687. In this depiction, the hero who killed Medusa wears the winged slippers of Hermes, which enable him to fly; he wears, too, a helmet which renders him unseen; and he wields a sword made of diamonds. A quartet of stars defines the face of the Gorgon: in this version, ρ Persei is at the tip of the nose; ω Persei is to the right of the nose; π Persei is the left eye; and β Persei, or Algol, is the right eye. The constellation, incidentally, is reversed left to right from the way you would see it if you looked at the sky. The reason is that some atlas-makers used a celestial globe in the preparation of their drawings. Hevelius, for one, did so. Accordingly, any given portion of the sky appears in this atlas as if it were inscribed upon the surface of a sphere and the observer were outside, looking downward upon it. As we actually look at the sky, however, it is as if we were *within* a sphere, and gazing upward.

by the Chinese, is called Rosh ha Satan (Satan's Head) and also Lilith by the Hebrews. There seems to be a general consensus in its naming—a consensus which is shared by the astrologers, who consider it the most dangerous star in the heavens: a carrier of misfortune, a mediator of violence. Probably its most revealing name is that of the Blinking Demon, but we ought not to count that title since Allen, writing in his popular book *Star Names and Their Meanings,* finds no evidence that it was known as such in antiquity.

Still, Algol is unique among all the stars in the sky in that it literally "blinks" in the course of a single night. This phenomenon was observed scientifically in the late seventeenth century, first by Montanari and then by Maraldi. But the most interesting communication about it, reporting the existence of minima and maxima in the brightness and an overall periodicity of two days and 21 hours, was done by a deaf and dumb 18-year-old boy and read at the Royal Society in 1782. The boy, John Goodricke, Jr., of York, died four years later, two weeks after being admitted to fellowship in the Society. By then he had discovered the variability of the stars β Lyrae and δ Cephei. He had also correctly guessed the source of Algol's variation, although Herschel never accepted the explanation.

Algol is an eclipsing binary star: a bright star with a dull companion that periodically gets in its way as seen from earth. For two and a half days, it remains more or less constant at 2.3 magnitude. But then, over about three and a half hours, it drops—first slowly, and then rapidly—to a magnitude of 3.5. It stays at this dimmer value for about two hours and then recovers to its fullest brightness over the next three and one-half hours. Of all the variable stars in the sky, Algol is not only among the brightest but is also the one that shows the steepest variation. Its time course of change is as shown in figure 6. Gorgonea Tertia is also a variable star, changing from 3.4 to 4.2 in magnitude, but not over so rapid a course.

One last astronomical fact must be mentioned here: the divergent point for the Perseids—the meteors of summer—lies at the wrist of Perseus. In other words, the wrist is the point from which all the meteor paths seem to issue; and this makes direct sense of a piece from the myth. Hermes, you will remember, gave Perseus a diamond sword. Why diamond? It was not then a cutting tool as it is now in electron microscopy. And I refuse to believe that significant words are tossed into a myth for the

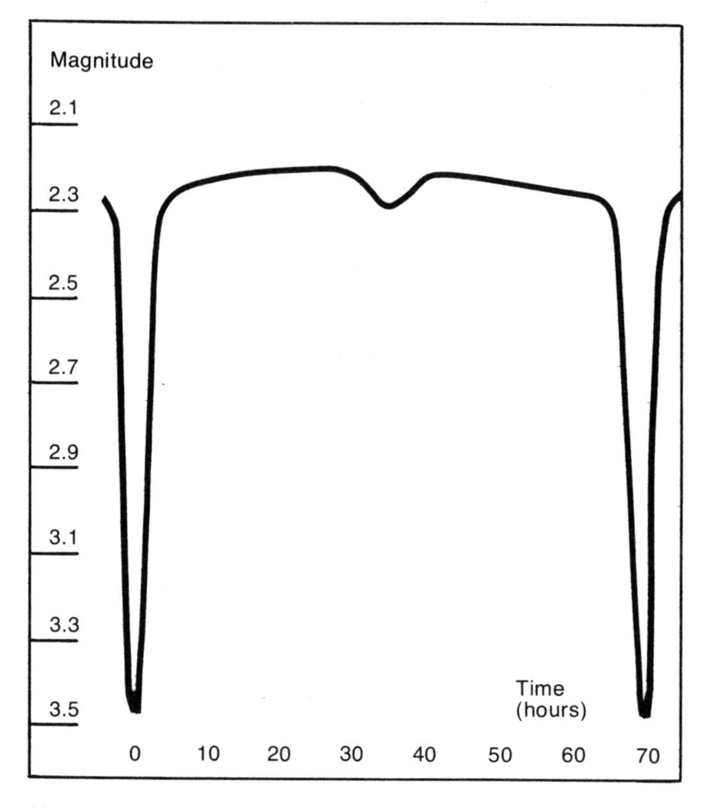

Figure 6
A light-curve for the star Algol. The vertical axis shows the apparent luminosity of the star; the horizontal axis shows time. Plainly, then, the star's brightness varies: it drops 1.2 magnitudes and then returns to its maximum value in a matter of hours. This dimming and brightening of Algol—this blinking, so to speak—occurs once every two days and twenty-one hours.

sake of mystification. The shower of Perseids in late July and early August occurs as if issuing from the wrist of Perseus the constellation, and I can imagine no more apt metaphor for that fan of falling stars than a sword of diamonds cleaving across the sky. But a strong feeling plus a coincidence do not constitute a proof, so I will only suggest that the myth is to be taken in this way.

Far more important is the Caput Medusae. To understand it properly, we are led to a mythic theme of great power and antiquity—the story of the Evil Eye. The best and most delightful account in English is Elworthy's book, supplemented in part by Gifford's later work, which relies heavily upon it.

The legend of the evil eye is as ancient as scribed artifacts. It existed already among the Sumerians and is well developed in the oldest records from Egypt. One finds it in China, India, the pre-Columbian cultures of the Americas, among the Eskimos, the natives of the Kalahari, the inhabitants of New Guinea, the Polynese—in short, everywhere. It is the eye whose glance brings evil. To be "fascinated" is to be ensorcelled by this eye.

Among the most amusing modern examples of the evil eye is the case of that amiable nineteenth-century pope, Pio Nono, whose malocchio was so notorious that workmen refused to renovate the Vatican unless the pope were guaranteed to be elsewhere. The quintessential evil eye, however, belongs to mythic characters. One of my favorites is the Irish hero, Balor, whose eyelids were swollen up in his youth by accidental contact with fumes from a witch's cauldron. Balor became the major arsenal in his nation's army. He would be positioned to face the oncoming enemy, and four assistants with fishhooks would raise the swollen lid until his baleful gaze was exposed to annihilate the attackers. He came to a bad end, though, for it took so long to lift his lid that an enterprising rock-thrower hit his eye before it was exposed and knocked it through the back of his head, whereupon its gaze desolated his own army.

Minor legends of this sort are the joy of story-tellers everywhere. But the major myths of the evil eye are always terrifying. They include the accounts of Lilith, the Medusa, the Strix, the Onocentaur, the Lamia, the Empusa, to mention but a few in the occidental tradition. Various animals also have the power of fascination—the serpent, the owl (who is the bird of Pallas), the cockatrix, the basilisk or salamander, the locust and the grasshopper, and so on. It is hardly surprising that from such a tradition the perceiving eye emerges as an emissive rather than a receptive organ.

The mythic figures from antiquity that are mentioned above share several common features: they are all feminine, they all have removable or alterable eyes, and they are all malignant. One might have imagined that things should have been otherwise—or at least that certain male figures, notably Argus and the Cyclops, should have shared the females' traits, for their mythic significance also rests on their eyes. Yet those latter orbs are neither powerful nor evil. Nor is the eye of Zeus himself, though he sees through the walls of Danae's prison. It seems that the comparative mythologists speak with reason when they talk of the Eye-Goddesses—including Ishtar and a whole host of figures from India and the orient generally. Although the concept is far from clear, and certainly far from general acceptance, the motif of all these goddesses is the powerful eye.

Consider Lilith. According to a Talmudic account, she was Adam's first wife and a daemon, but not necessarily malign at first. She was a night spirit, and bore Adam's children before Eve was brought forth from Adam's rib. Then, however, she was abandoned for Eve, whereupon she destroyed her own children and went forth as one of the midnight hags to prey on the children of others. She fascinated; and she could remove her eye and hide it. To her is attributed both beauty and a terrifying appearance, quite like that of the Medusa. In a later period she merges with the Lamia who preys on young men as well as children. The newborn, in particular, had to be protected from such deadly ones.

It is interesting, in this respect, to note that the newborn infant in Greek villages, even at this century, is protected specifically from starlight for at least eight days and up to a month and a half, for otherwise the stars might "overlook," i.e., fascinate, the child.

Now language is always a great trap in all texts, sayings, and doctrines that come down from the past. This is doubly the case when one deals with the oral tradition, in which nothing is written down. It is therefore of only passing interest that one of the bucolic English phrases for being affected by the evil eye is "to be blinked over," and that the English name for Algol is "the Blinking Demon." I must lay ground far more carefully.

The thesis that I mean to advance is that Algol and quite possibly Gorgonea Tertia (ρ Persei) were the eyes of the Gorgon, of Lilith, of the Lamia, of the creatures whose glance was deadly. These stars became the evil eye in the sky particularly by virtue of their blinking; and they were thus the nucleus around which the constellation Perseus was erected. I mean also to suggest that the constellation Perseus is one of the longest in the sky because the "diamond sword" had to be accommodated since it, too, was associated with the death of infants. And, finally, we know these features, I propose, precisely because there is no direct evidence for them.

Let me pursue, first, some astronomical matters. Algol, as I remarked, blinks quite visibly in the course of some nights. One does not need a telescope or any other instrument to tell that it varies, for it changes almost a whole magnitude in but three hours, and given the stars around it for comparison (especially Persei ρ and π), the change is eminently noticeable to watchers such as young Goodricke. Accordingly, given that the star-watching of antiquity was not always directed at the horizon, for the zodiac and the constellations had to be defined, and given the appreciation of secular process that identified the planets—the wanderers—and even the slow precession of the skies from year to year, is it reasonable to suppose that a bright, blinking star could escape notice? One might argue that all the stars were eyes in the skies, yet this, I suspect, is a great folly. What characterizes the eye of all land animals as well as the Medusa herself is the veiling and unveiling, the opening and shutting. It would be a surprising coincidence that Algol defines the head of the Medusa had the blink not been noticed.

But then why is Algol never *said* to be the eye? Heaven knows that the astronomers and astrologers are specific enough about forearms, elbows, knees, and a variety of other parts. Indeed, the various parts of Perseus himself are identified, and furthermore, the four Gorgoneae are plainly held to be a head, the Caput Medusae, by Hipparchus and Ptolemy. It is thus a bothersome omission that the most dangerous star in the sky is not identified except by indirection—or rather, it is bothersome until one recollects, with some poignancy, the nature of naming.

After all, it is not obscure or newly found knowledge or a mere invention to point out that if one named a demon or a malign influence, one attracted its attention. Remember how the Eumenides (the Furies of classical mythology) were never named as such, but called instead the Gracious Ones. Remember the proscriptions directed by the Jews against the utterance in any way of the true name of God. From Ra to Rumpelstiltskin, the names, if uttered, generate a direct connection of the power to the self. And this precept, strong enough in religious matters, becomes of even greater consequence in daily beliefs, or superstitions. In the hands of the elite, naming gave a touchy power over demons. And in the mouths of the other classes, naming was fatal. This is a basic social rule.

Here, then, is a demon's head among the stars, its eyes blinking down on the world. Who is intrepid enough to call forth the "overlooking" (the supervision, the episcopacy) of that eye upon himself by naming it, describing it, examining it carefully to determine the rapidity of blinks—in other words, by doing everything possible to call its attention?

Now let me add a trivium. What do falling stars signify? Does it really surprise anyone to know that across many cultures each meteor corresponds to a sinless soul entering heaven, to a newborn child dying, to the death of a saintly one—always to death? But this shower of Perseids is, I claim, the diamond sword wielded by the hero who carries the Gorgon's head whose blinking eyes covet the souls of the newborn—the Lilith, the dread Lamia, the Cacodaemon.

We are told from ancient times that the sea and the sky are mirrors of one another. We see now that the mirror is not meant in a trivial sense; it is a mirror of relations. That is to say, what characterizes one can be read into the other, and without such a reading there is no language for expressing an astronomy in the first place.

This must be true even for physical reasons. Suppose you were to maintain that the ancients meant a literal mirroring of the sky in the sea, pointing out the lovely symmetry of the moon and its reflection on a calm night. But have you ever seen the *stars* reflected, save perhaps Sirius? For water, as glassily quiet as in a windless lagoon, drops what it reflects by four magnitudes, as Minnaert remarks. Thus we can guess why Perseus comes to the Medusa with reflecting shield—for what you can't see in a mirror can't see you in that mirror. Think, if you will, of stars reflected on a polished brass shield and tell me—given the specular and diffuse reflectances of brass, can you see Algol or, indeed, the eye of anyone behind you? I think it possible that there never existed, during the eras in which the Perseus myth formed, a mirror capable of showing the stars, and that is the meaning of our hero's shield. (Narcissus fascinated himself in daylight and on still waters in a glade because it is just possible to make out your own eye against the bright background of the sky light, even filtered and reflected through the trees.)

It seems to me that the study of archaeoastronomy now proceeds along two lines. On the one hand is the thoroughly admirable strategy of looking at ancient artifacts and deciding what they signified. On the other hand is the thoroughly dangerous strategy of exploring the poetical language of myth. It is dangerous, for if all you can see in the myth is the human unconscious—if all you can see is yourself—then you have embarked on a second-rate venture. But if you can see the stars in a myth, that is far better. For you don't know the stars as well as you think you know yourself.

It is dangerous, too, because you are sure to go too far when you attempt to reinterpret myth. I myself have gone much too far in this article, and so can be accused of making my own myth to render memorable the sundry places in the sky. I am not offended by that charge at all. If you find the story of the octopus, Algol, and the Gorgon Medusa irritating enough to recall, I will have explained the ancient arts of memory more by illustration than by proof. Most of you, of course, may prefer a rational account of things; but I was never one to put Descartes before Horus.

For further reading:

The Evil Eye, F. T. Elworthy, Julian Press, New York, 1953.

The Nature of Light and Colour in the Open Air, M. Minnaert, Dover Publications, New York, 1954.

HARALD A. T. REICHE

THE LANGUAGE OF ARCHAIC ASTRONOMY: A CLUE TO THE ATLANTIS MYTH?

It is clear that Stone Age man lacked our coordinate systems, our spherical geometry, and our computational techniques. Indeed, it is clear that he lacked writing itself. It is equally clear, however, that the administrators in charge of Stonehenge and similar structures succeeded at the following exacting tasks: they justified to a larger lay public the great communal effort required for the construction of the facilities; they explained to their assistants the rationale and the procedures for working with the numbers and celestial alignments embodied within; and they insured, again without writing or mathematical notation, the accurate transmission of these procedures from generation to generation across the centuries. How did they do it? What did they use for the requisite technical language?

In attempting to answer this question, we are the beneficiaries of a line of investigation unrelated to the decoding of Stonehenge and similar efforts by investigators such as Alexander Thom and Gerald Hawkins. First of all, in the late nineteen twenties, Milman Parry demonstrated at work in the language and structure of the Homeric epic a technique of composition still discernible in contemporary Serbo-Croation oral poetry. The crux of the technique is a vast storehouse of formulaic phrases ("rosy-fingered dawn," "swiftfooted Achilles," . . .) and even entire formulaic lines ("But they raised their hands to the foods spread out before them.") collected and recollected by many generations of practicing bards. By means of them, whole sequences of events in epic recitals can be remembered, for the phrases and lines are fixed, "prefabricated" elements, at once metrically correct and positionally predictable. In much the same fashion do the left hand's familiar combinations of chords enable a pianist to remember a given melodic line.

The appearance of formulaic devices in Homeric composition strongly suggests that the form in which Neolithic and Bronze Age astronomers explained and transmitted their knowledge was rhythmic (i.e., metric and versified) speech, perhaps coupled with melody and (to judge by analogy with Far Eastern temple dances) elaborately stylized pantomime. After all, a useful technique of memorization is unlikely to have remained undiscovered until quite recently. It has even been suggested, plausibly it seems, that Agamemnon may have cast in verse form his instructions to the fleet assembled at Aulis for the expedition against Troy.

Then there is the work of Hertha von Dechend, historian of science at Frankfurt University. For the past twenty years Professor von Dechend has been comparing and assembling the pieces of an immense puzzle composed of specific and recurrent tales, formulae, and motifs drawn from the mythology of virtually the entire globe. Her evidence overwhelmingly argues that in the preliterate, archaic cultures that coined them, these elements functioned as vehicles at once for expressing, for memorizing, and for transmitting certain kinds of astronomical and cosmological information. This should not amaze us. Consider that in our own living memory, a "story told" (the English equivalent of the Greek word *mythos*) enabled Polynesian navigators without compass, chronometer, and maps to effect pinpoint landfalls hundreds of miles away, for such a story, by way of a fictional narrative, connected the stars that, from a boat crossing the Pacific in the proper direction, are seen to rise (or set) at the same horizon point. In a similar fashion, according to von Dechend, it is in the form of so-called etiological or explanatory myths, almost certainly versified (so it is here suggested), that early man acknowledged, and hypothetically sought to account for, certain flagrant violations of spatiotemporal parsimony and symmetry in the sky. Plainly, such violations troubled not only ancient observers; witness the complaint of King Alfonso X of Castille, patron of planetary tables published in 1273 A.D., that if God had but consulted him at creation he could have given Him some good advice.

Almost invariably (to judge from von Dechend's research), it is monstrous deeds—mutilations, incest, cannibalism, all of these inflicted by celestial beings upon their fellows—which keen observers felt logically driven to postulate as (at least generically) adequate explanations of discrepancies in the sky. Item: In many ancient civilizations, the inclination of the sun's (and the planets') path with respect to the planes in which the fixed stars diurnally revolve was taken as prima facie evidence of some earlier celestial disaster, much as we take a broken windshield on an automobile as evidence of some prior accident. Read in this context, the story that Kronos (Saturn) castrated his father Ouranos so as to separate "sky" from "earth" is plausibly taken as an attempt to account for the obliquity of the ecliptic (the yearly path of the sun) relative to the celestial equator. Similar stories were current in the ancient Near East, whose influence upon Hesiod, our earliest source for the Greek castration myth, is now generally recognized.

Item: The incommensurability of the solar and lunar years was explained in Egypt as the deliberate departure from an original 360-day cycle of twelve 30-day lunar months. In a game of dice, so the story goes, Mercury won five days from the lunar year and added them to the solar year (resulting in the present values of 355 and 365 days respectively). By doing this, he created the extra days on which the planetary quintuplets could be born. For those children had been conceived in an incestuous union by the sky goddess Nut, and the sun had prohibited their birth "on any day of the month or year."

Item: Consider the slow eastward slippage, past a fixed and ancient horizon marker, of the familiar constellations marking solstices or equinoxes, clearly noticeable after but a few generations and entailing the gradual obsolescence of any given polar star. (The illustrations in figure 1 demonstrate the effect.) All one needed to notice this (in Professor Philip Morrison's apt phrase) was an old tree and faith in the veracity of one's grandfather. Obviously, a set of stone markers of the sort amply documented by Professor Alexander Thom's work would have done even better. Given these naked-eye horizon phenomena, there is no empirical reason for down-dating the recognition of the slippage in question to the time (150 B.C.) when Hipparchus correctly traced the visible phenomenon at issue to the westward "precession" of the invisible intersection of the invisible celestial equator with the equally invisible ecliptic. After all, the correct explanation of a phenomenon commonly comes long after its routine observation. And for that matter even Hipparchus

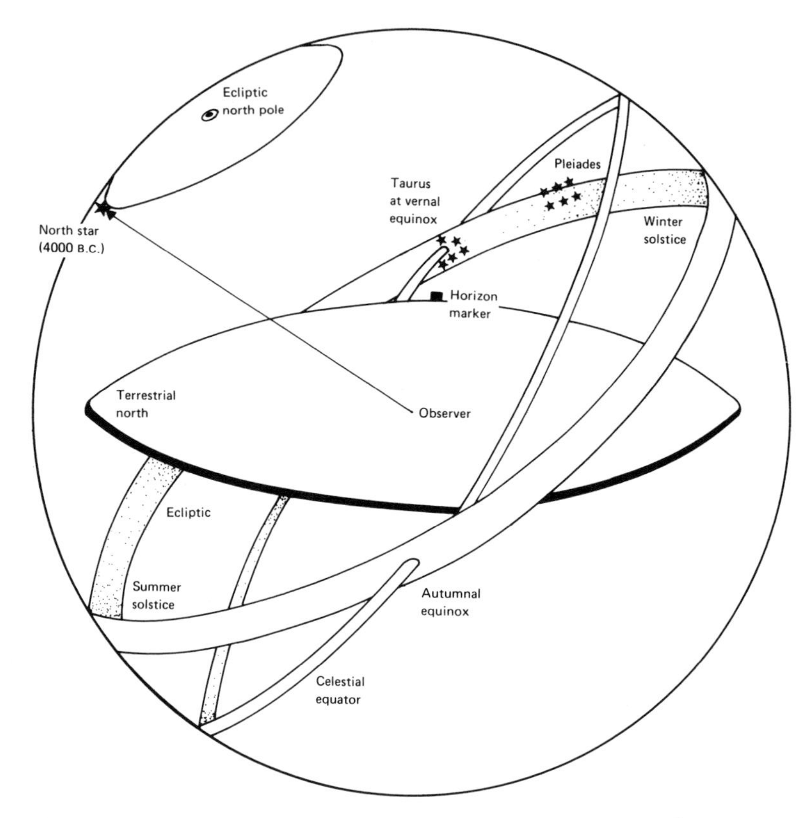

Figure 1
The slow eastward slippage of the zodiacal constellations, together with our modern understanding of what causes the phenomenon. The celestial sphere at the left shows in schematic form the relations that obtained in 4000 B.C.; the angles are those applying for an observer at the latitude of Babylon—about 32.5 degrees North. On the day of vernal equinox, the observer establishes a marker at the horizon to show the place where the sun rises. As the stars disappear in the morning light, he observes that the sun occupies the house of the constellation Taurus. The illustration also shows the Pleiades, higher in the sky. The celestial sphere at the right shows, again schematically, the alignments that obtain for a Babylonian observer 2,800 years later, in 1200 B.C. On the day of vernal equinox, he finds that the sun occupies the Pleiades, and that he must move the ancient horizon marker to the west in order to accommodate this fact. Taurus, meanwhile, has sunk in the sky on an eastward slant, and is only partially visible at the horizon; the observer of 1200 B.C. might say that it is "drowning," and in fact the ancients appear to have described it in that fashion. The *modern* understanding of such changes in the sky is that the axis of the earth's rotation is performing a twenty-six-thousand-year circle around the axis of the sun's yearly path against the background of the zodiac. (In the illustrations, the first of these axes is a

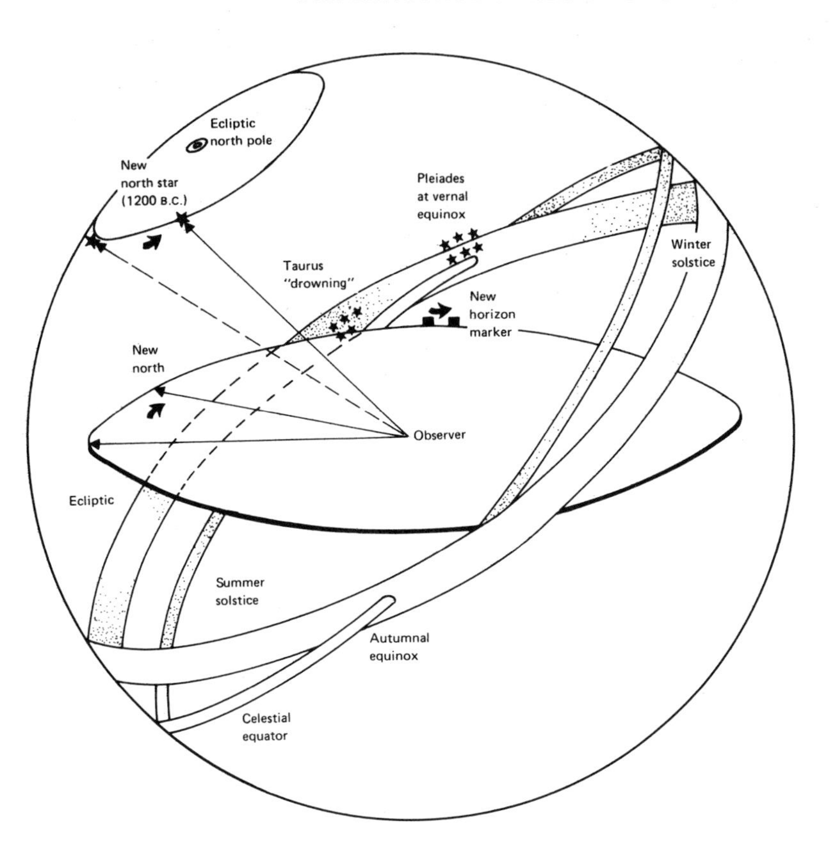

line directed from the center of the celestial sphere—the point marked "observer"—to the point marked "north star"; the second would be a line directed from the observer to the "ecliptic north pole." Comparing the two spheres will show that the former is indeed circling the latter.) The cause of this so-called precession is a torque on the earth due to the gravitational field of the sun; but in any event, the result of the precession is the phenomenon at issue, for the precession causes a westward drift to the intersections between the "ecliptic" and the "celestial equator"—the gyroscopes, if you will, whose axes we have just pointed out. Note from the illustrations that these drifting intersections define the autumnal and vernal equinoxes.

failed to provide the correct kinematic explanation—an explanation not furnished till Newton.

There remains the standard objection that, had the "precession of the equinoxes" been recognized by pre-Hipparchian astronomers, Plato would have allowed in his models for a second and slower westward motion of the celestial equator, additional to the diurnal one, and Eudoxus would have provided an extra rotating sphere to handle it. (Eudoxus attempted to model the planetary motions on a set of concentric spheres within the larger sphere of the universe itself.) This might be true if, as is not the case, the precessional phenomena had been correctly construed as cyclical, i.e., as due to the westward advance along the ecliptic of the point where the equator intersects the ecliptic. To judge by Plato's *Politicus,* however, the phenomena in question were construed as alternating or reciprocating, as if, to use a modern analogy, they were governed by a spring-driven mechanism, alternately wound up and running down.

Plato's omission in the *Politicus* to specify a rate of change for this stellar drift, indeed its total omission from the polyspherical models of the cosmos of both Plato and Eudoxus, and Eudoxus' combination of an excellent value for the tropical year with inaccurate locations of the equinoctial points suggest the following conclusions. Plato, and probably Eudoxus too, learned of the phenomena at issue not through personal observation but at second hand. They learned of them from a source that happened to omit all kinematic and quantitative specifications, i.e., presumably through Egyptian "myths" like the ones cited by Plato (*Timaeus* 22; *Critias* 112A). Eudoxus, in particular, learned of them from a source unrelated to the Babylonian sources from which he demonstrably derived his conventional and empirically inaccurate locations of the equinoctial points (at 22° of their signs in an early work and 15° in a later work). Finally, both Plato and Eudoxus felt sufficiently sanguine about the kinematic axiom of uniform circular motion and sufficiently diffident about the empirical details of "precessional" motion to be prepared to sacrifice the latter on the altar of the former—at least pending sharper information and/or the availability of a better axiom. After all, this sort of tug-of-war between theoretical models and notorious phenomena is (as Saltzer has recently stressed) a central and well documented aspect of the whole history of Greek astronomy. Witness Ptolemy's inability, within his geometric model, to do more than register the retrograde motions of the interior planets; and Eudoxus' inability, within his homocentric spheres model, to accommodate such notorious phenomena as the variable size and brightness of all planets and even such prominent retrogradations as those of Mars and Venus.

Now Professor von Dechend has succeeded in identifying a number of mythological formulae and motifs which, if they are not patronizingly to be dismissed as the confused phantasies of childlike "primitives," make sense as hypothetical explanations of precisely this slow eastward drift of the solstitial and equinoctial constellations and the consequent periodic obsolescence of those cardinal points by reference to which we orient ourselves here on earth.

Take the story of how at the end of the so-called Golden Age the sun was displaced from his former path by the fact that Phaethon, the sun god's son, lost control of the solar chariot; or the story of how Zeus, disgusted by the cannibalistic dish fed him by Lykaon (who wished to test Zeus's omniscience), kicked over the table in horror. Both stories, so von Dechend shows, refer to successive phases of one and the same process, the gradual displacement of the familiar constellations marking vernal and autumnal equinox and summer and winter solstice by new constellations to the west of them. Thus Phaethon's Fall marked the end of the Golden Age, the great world age when the Milky Way, linking Gemini and Sagittarius, had coincided with the sun's equinoctial place. The episode of Lykaon's dish marked the end of the next or Silver Age, when Taurus and Scorpio had replaced Gemini and Sagittarius as the constellations governing the equinoxes (and the north pole had, of course, shifted accordingly).

To be sure, the mythological language describing the eastward slippage of familiar equinoctial and solstitial constellations originally did so in terms of the corresponding naked-eye phenomena only, i.e., not in terms of the equator-ecliptic coordinate system, which, as we have noted, is a relatively late invention. Thus a constellation hitherto heralding sunrise at vernal equinox would seem to rise ever earlier with respect to sunrise until it had slipped so far eastward beneath the eastern horizon as to be still invisible at sunrise. In horizon terms, this might be spoken of as "drowning" or "being drowned"; in ecliptical terms, it would be "climbing onto a mountain or into the sky from the shore" (i.e., to a place in the zodiac just above the equator).

Meanwhile its replacement would, in horizon terms, descend from the sky to be the new harbinger of sunrise at vernal equinox; or, in ecliptical terms, "climb from the waters below (sc., the equator) onto the dry land" marked by the equinoctial point. At sunset on the day of the vernal equinox, the old constellation first visible above the place where the sun set (i.e., the same which heralded sunrise in the morning) will have correspondingly shifted eastward into the sky until it has become invisible in the light of a sun still well above the western horizon. In ecliptical terms, it would be "drowned" or "drowning" in the watery section of the zodiac beneath the equator. And the new constellation marking the place of sunset would, in horizon terms, have "climbed" into visibility from the "waters below"; or, in ecliptical terms, "descended from the sky or a mountain to the seashore." (Again consult figure 1.) Note that the replacement of the horizon by the ecliptic-equator reference system merely involved a reversal of directions. What for the old and new constellations heralding sunrise was, in horizon terms, drowning and descent from the sky respectively, now became, in ecliptical terms, climbing into the sky and onto dry land respectively. And what for the old and new constellations marking the place of sunset was, in horizon terms, rising into the sky and onto dry land respectively, became, in ecliptical terms, drowning and descending from sky to seashore respectively.

It is neither possible nor necessary here to rehearse these findings in detail. For the massive infrastructure of supporting data the reader is referred to Professor von Dechend's publications already available and forthcoming. Suffice it here to single out four corollaries only:

- The mythological motif of successive world ages to denote successive equinoctial constellations (not, of course, as witnessed by any one generation, but as an accumulated tradition or as inferred by simple backward extrapolation).
- The mythological motif of monstrous deeds followed by "catastrophies," usually floods or fires or both, opening and closing each of the world ages.
- The association of each world age with a planetary ruler, i.e., with one of the naked-eye planets, in an order varying inversely with the angular velocities of the planets across the sky. In particular, the radial distances of the planets were taken to be inversely proportional to their angular velocities. Thus, Saturn,

the slowest of the planets, was taken to be the most remote, and also the oldest. And by this reasoning, Saturn was construed to have been the ruler of the most ancient world age, the Golden (the planet having been displaced outward to its present position by the planetary rulers of subsequent ages).

• Finally, the consensus, virtually universal among ancients as different as the author(s) of the Babylonian creation epic and the far later writers Solon, Herodotus, Caesar, Plutarch, and Tacitus, that the "true" identity of a foreign deity can be deduced from its attributes. Thus, the deities in charge of the upper, middle, and lower portions of the sky and those in charge of the several naked-eye planets in one culture are almost always unhesitatingly translated into those familiar to another. Witness, for example, the functional identity of what the Babylonian epic terms the Ways of Enlil, Anu, and Ea with what Greek mythology distinguishes as the realms of Zeus, Poseidon, and Hades—namely the upper, middle, and lower sky, or, in a metaphorical variant, the bright vault of the sky; the earth and its supporting waters; and the underworld, the unseen bottom of the sky. Witness, too, the functional identity of the Babylonian god Enki-Ea with the Greek Kronos and the Roman Saturn. Solon's "translation" of deities mentioned to him by Egyptian priests will presently prove to be a case in point.

In the remainder of this article, I propose to reconsider Plato's Atlantis myth in the light of these four corollaries. The story itself is distributed by Plato over two dialogues, the *Timaeus* and the *Critias,* both set in 421 B.C. (when Plato himself would have been seven). In the first, the grandfather of Plato's maternal cousin (both are named Critias) tells the story in summary form, by way of a preamble to Timaeus's long account of cosmology, astronomy, physics, and biology, all of this cast in the form of a cosmogonic myth. Critias (Plato's cousin's grandfather) explains that the story of Atlantis reached him by word of mouth from *his* ancestor, Solon, and that Solon had heard it (probably, one would guess, soon after 570 B.C.) from Egyptian priests at Saïs, in the Nile Delta. It has been claimed, we should note, that these details about the pedigree of the story constitute a mere literary device, meant to get us to accept a myth (a "probable story": *eikos logos*) that Plato himself has invented. But why should Plato have hoped to deceive his readers in that way? Athens was small, Plato's family was large, and Plato was an aristocrat with a reputation for veracity to protect. We have good reason, then, to accept Plato's assurances that Solon heard the story in Egypt.

In the second dialogue, Critias is the only speaker. He tells in detail what prehistoric Athens and Atlantis were like, but then unaccountably breaks off his narrative.

Very briefly, the story is this: Nine thousand years before Solon's visit to Saïs, the gods took charge of and peopled their appropriate allotments of the earth. Athena and Hephaestus jointly got Athens; Poseidon (or at least the Egyptian god whom Solon equates with the Greek "Poseidon") got the large island of Atlantis—an island beyond the pillars of Hercules, vulgarly taken to be the Straits of Gibraltar but even in that meaning suggesting the limits of the known world (figure 2). A millennium later, Athena got Saïs. To isolate and protect the mountaintop on which he mated with a mortal woman on the island of Atlantis, Poseidon created an island within an island by ring-

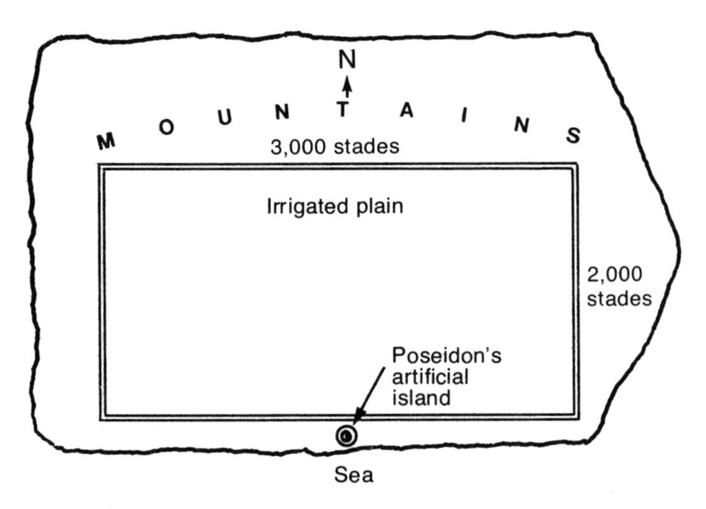

Figure 2
The island of Atlantis, bestowed upon Poseidon (according to Plato) when the gods apportioned heaven and earth among themselves at the end of the Golden Age. The island includes an irrigated plain—irrigated, at least, by Poseidon's descendants—measuring 3,000 by 2,000 "stades" (a state is 606.75 feet). Poseidon himself, to protect the place where he mated with a mortal woman, constructed on Atlantis a sanctuary within five concentric bands alternately of earth and water. This island-within-an-island appears at the bottom of this large-scale drawing, and in greater detail in figure 7. One further detail: the eastern side of Atlantis comes to a tip—a tip said by Plato to point toward the "pillars of Hercules."

ing the mountain with five concentric bands, two of them moats. After his five pairs of twin sons had come of age, he distributed the larger island among nine of them but reserved his artificial island and the belt of land surrounding it, which was largest and best, for his firstborn son, Atlas (who in Egyptian as well as Greek mythology kept sky and earth apart). Poseidon made him king over the other nine "rulers" (*archontes*); and at that point Poseidon presumably withdrew, because no more is said about him. After "many generations" the governance of Atlantis developed the nation into a vast naval empire ruling the lands up to (but presumably not initially including) Egypt and Tuscany. Trade, natural fertility, and the mining of metals combined to produce enormous wealth, which enabled Atlas's descendants to add an extra ring of land to the central island, to build walls around each of the four resulting circular areas of land, to run a navigable canal straight from the sea to the central island, to build a costly temple to Poseidon in the center, and to develop into an enormous, irrigated rectangle the arable portion of the larger island. The ten ruling princes also convened assemblies in Poseidon's sanctuary at five- and six-year intervals. On these occasions they hunted a sacred sacrificial bull, armed only with wooden sticks and nooses.

"Many generations later," however, as their admixture of divinity grew ever feebler, the rulers of Atlantis succumbed to lawless ambition and power. Whence Zeus decided to teach them a lesson so that they would mend their ways. Here Plato's *Critias* breaks off. But we know from earlier hints (*Timaeus* 25 B, D) that two events then occurred. Trying to capture the whole Mediterranean in a single eastward sweep, the Atlantid forces were defeated by prehistoric Athens in a war so long that it involved at least four successive mythical kings of early Athens (Cecrops, Erechtheus, Erichthonius, and Erysichthon [*Critias* 110A]). So much, presumably, for Zeus's attempt to bring the ten rulers of Atlantis to their senses. That was 9,000 years prior to Solon. At an indefinitely later time, however—the lesson presumably not having been learned—Zeus sent (or authorized Poseidon to send) earthquakes and floods, which in one terrible day and night drowned not only the entire island of Atlantis, rulers and all, but also the occupying Athenian army and navy (*Timaeus* 25D). At the same time torrential rains swept away much of the soil, and so altered the climate, of prehistoric Attica (*Critias* 112A).

The history of the interpretations of this myth can be arranged along a spectrum. At one end (1) are those who claim that the story is literally true; at the other end (4) are those who claim that it is either a purely decorative fiction or one that, at least in its original form, reflects Egypt's foreign-political aim of securing an Athenian alliance against Persia. Between these two extremes fall two intermediary positions. First, there are (2) those who, besides crediting the myth with historical truth, take it to symbolize in addition the allegedly timeless truth of Plato's distinction between a finite set of fixed, purely intelligible archetypes underlying all of reality, on the one hand, and the infinite, ever fluctuating variety of their physical embodiments, on the other (so the ancient Neoplatonists Iamblichus, Syrianus, Proclus). For the historical basis, the advocates of position (2) seek a geographic referent (2a) outside Gibraltar (anywhere from Spitsbergen to America) or (2b) inside the Mediterranean (e.g., on the site of the present, approximately circular, Santorin group of islands). Then (3) there are those who, while denying any historical basis, see the story as an allegory (3a) of the nature (astronomical or eschatological or both) of the universe, or (3b) as a fictional projection into prehistory of what in the *Republic* was described by Plato as utopia.

Those favoring a location for Atlantis outside Gibraltar (2a) can appeal to Plato's express language (*Timaeus* 24E). Those favoring a location inside the Mediterranean (2b) can appeal to the historical seaquake which about 1200 B.C. destroyed Crete, if at the price of the auxiliary postulate that Solon (or his informants), in translating the Egyptian numerals concerning date and size of Atlantis, mistakenly inflated them by a factor of ten. Finally, those favoring only a fictional, utopian interpretation for Atlantis (3b) can cite what by Plato's own admission (*Timaeus* 26E) is the "strange parallelism" between the utopian Republic and early Atlantis.

Several of these intermediary views are compatible, of course. Thus, an Egyptian cosmological account, originally encoded in story-form and involving "Athens," could well have been furnished Solon for a political reason additional to that of impressing him with Egypt's hoary antiquity. And Plato, in editing this tale, could perfectly well have chosen to suppress some aspects of the story in favor of political, and above all cosmological, ideas of his own. With these modifications, Alexander von Humboldt's view that Plato quite possibly made use of native Egyptian myths retains its plausibility. In any event, the traumatic folk-memory of the seismic catastrophe of 1200 B.C. may well have colored Plato's description of Atlantis's end—without the latter catastrophe being reducible to the former.

That granted, the point of the following pages is to call attention to a body of evidence which argues that, among much else, we have here an etiological myth concerning the succession of astronomical "world ages."

The doctrine of the world ages occurs as early as the Babylonian *Enuma Elish* ("When above . . ."), dated to the middle of the second millennium B.C. It has analogues in Iran, Greece, Rome, the Norse Edda, and elsewhere. We have already had occasion to note that it represents the slow eastward drift of equinoctial constellations in the sky. Consider now the mythological explanation devised to account for the role of so conspicuous a swath of sky as the Milky Way, linking Gemini and Sagittarius by way of a luminous bridge. A simple backward extrapolation would have sufficed for the ancients to have established a period in the still more distant past (from about 6000 B.C. through 4000 B.C.) when this bridge still coincided with the sun's place at dawn and dusk on the equinoctial days of the year, i.e., when the equinoctial sun would have been seen to rise and set in the constellation of Gemini in the spring and Sagittarius in the fall.

What, in retrospect, made this period appear "golden" was that gods and men were believed to have communed by way of the bridge. Accordingly, when the Milky Way had ceased (about 4000 B.C.) to overarch the equinoctial points, this was interpreted as the end of the Golden Age and, by that token, as the start of a steady decline of which the calamities of the ancients' own age, that of Iron, were but the ultimate consequences.

The Golden Age was the period in which the universe acquired its initial shape. Consistently, therefore, the planetary ruler of the Golden Age—namely Saturn (Kronos in Greek)—is in virtually all mythologies conceived as the Engineering God (*Deus Faber*), responsible not only for the fathering of his planetary successors but also for the original design and realization of cosmic order. Now aside from genealogically linking the younger to the older gods, the myth-makers among the ancients had treated each world age as self-contained and as opening and closing with some catastrophe. Plato, however, shared the Pythagoreans' commitment to strict logical coherence and its corollary, a purely deductive model of truth and reality. He therefore retained the myth of world ages as introduced by a Saturnian Demiurge, but recast it in such a way that what formerly had been the products of random, if successive, catastrophes now appeared as the necessary consequences of what had been implicit in the initial conditions. Thus, although, or precisely because, in the *Timaeus* the basic frame of the universe is traced to a divine engineer possessing all the earmarks of Saturn, the obliquity of the ecliptic is presented not as the result of a struggle (much less a castration, as we mentioned earlier) but as itself part of the initial design.

To this extent, then, Plato is undoubtedly original when in the *Timaeus* he has the Demiurge peacefully engineer what we understand as the equator-ecliptic coordinate system. At the same time, however, the notion of an Engineering God is demonstrably part of an older, apparently world-wide, tradition. Thus the god of the *Critias* whom Solon rendered as "Poseidon"—the Engineering God (figure 3) who works elaborate improvements on the mysterious island of Atlantis—evokes specific parallels. Indeed, Solon must have been faced with numerous possibilities in attempting to assign a Greek identity to the god of Atlantis. The essence of his difficulties no doubt lies in the multiple meanings traditionally associated with the term "water." For the sky was notoriously held to fuse with the sea at the horizon, whence the Greek *okeanos* encompasses both. Consequently, the term "waters" could mean one or more of the following: it could be the sea as familiar to a nation of sailors and fishermen; it could be the visible sky above our heads (the "waters above" of *Genesis* 1:7); or it could be the invisible sky beneath (the "waters below" of *Exodus* 20:4). Upon the banishment of Kronos/Saturn at the end of the Golden Age, Poseidon received the first of these, Zeus the second, and Hades the third. Complicating this problem is a further linguistic convention of archaic astronomy: the half of the ecliptic extending beneath the celestial equator was termed "wet" or "water"; that extending above the equator was termed "dry" or "earth"; and by this token, westward motion along the ecliptic carries one into "the waters below." Such, in fact, is the route taken by the souls of the dead on their voyage to the underworld; and such is the route which someone like Odysseus must take, even while living, if he wishes to visit there. At the southernmost point of the ecliptic (i.e., at the point on the ecliptic deepest into the "waters")

Figure 3
Poseidon reaches downward with a set of luminiferous calipers to map out concentric bands of earth and water on the island of Atlantis. The illustration, made by William Blake in watercolor, black ink, and gold paint, has come to be called "God Creating the Universe," but it might with equal justice be taken to represent the mythological theme of an Engineering God designing the cosmos. In Greek mythology, Kronos creates the design of the universe at large; and in Plato's writings, Poseidon, acting in his role as Engineering God, creates the design of his island.

lies the constellation marking winter solstice. This is the portal to the underworld. Upon arriving there, souls and visitors alike must presumably turn south to reach their destination in the unseen sky.

Recall, now, that the island of Atlantis lies "in front," i.e., outside or to the "west," of the pillars of Hercules (*Timaeus* 24E). Quite plausibly, then, that island is Poseidon's. For westward motion on earth, or on the ecliptic, places the island in a watery realm. Of course, the interpretation of "westward" in celestial terms requires that the "pillars of Hercules" be taken to refer not to Gibraltar, but instead to some suitable celestial analog, say the equinoctial points or perhaps the tips of the horns of Taurus. However, there are other possibilities for Atlantis. Westward along the ecliptic is also southward, and hence an excursion into the area normally associated with Hades. On that view the island would be in *his* realm. There is, moreover, the fixed association of Kronos/Saturn, at the close of the Golden Age, with banishment to an "island" at or near the celestial south pole, i.e., at a specific place within the underworld—an island on which Kronos sleeps while time stands still. It is interesting to note that Enki-Ea, the Babylonian Kronos, dwelt in the "town of Eridu at the confluence of rivers"—a place which remained a geographic mystery until, a generation ago, Van der Waerden could demonstrate the identity of Eridu with Canopus, the bright star in the southern sky nearest to the southern pole of the ecliptic. More about that later. The point for now is that the island of Atlantis could plausibly be the same as the island to which Kronos/Saturn is traditionally relegated.

One further association is possible. Atlas's island is reported by Hesiod (*Theogony* 517-20) to be to the west (near the Hesperidae), and by both Homer and Hesiod to be the place where Atlas upholds the sky, either by guarding the cosmic axis (so Homer) or by supporting it on his shoulders (so Hesiod). On either view, the island of Atlantis would again be at or near the celestial south pole—the more so if we can equate the island of Atlas with that of his daughter Calypso, for the latter island is expressly said by Homer (*Odyssey* I 50-54) to be "at the navel of the sea," i.e., at the deepest point of the "waters."

One sees, then, that to a Greek, the Egyptian deity responsible for Atlantis was bound to combine aspects of Poseidon (lord of the sea, and indeed dwelling in its depths [*Iliad* XIII 21]), Hades (lord of the underworld), Kronos (dwelling upon an island where time stands still), or Atlas (dwelling, it seems, at "the navel of the sea"). Under the circumstances, Solon may be pardoned for provisionally settling upon Poseidon. Yet if, as here suggested, the Atlantid "Poseidon" is also a variant of Kronos/Saturn in his role as Engineering God, then the doctrine of the world ages, detailing among other things the accession and relegation of Kronos, is very likely to have left its traces on the story of Atlantis.

That is in fact the case. On the testimony of the Egyptian priests themselves, the flood that destroyed Atlantis was occasioned by a "displacement of the heavenly bodies from their course" similar to that which they claimed underlay the story of Phaethon's age-ending accident with the solar chariot (*Timaeus* 22). More specifically still, the priests identified (in *Critias* 112A) the flood at issue with "the third one prior to the Deucalian one" (so named because Deucalion survived it, Noah-like, by means of an ark). Now there are two conflicting traditions concerning the date of the Deucalian flood. One puts it at the end of the third or Copper Age, another in the Iron Age. On the former dating, the flood that destroyed Atlantis would be the one that closed the Golden Age. But this is unlikely both because the gods did not receive their allotments of the earth till after the close of the Golden Age, and because the gradual corruption of the Atlantians (which is expressly said to have preceded, both chronologically and causally, their destruction by drowning) would have been incompatible with the Golden Age. On the latter dating, the flood that destroyed Atlantis would be the one closing the Copper Age and so, presumably, would be identical with the one whose sole survivor, Dardanus, was alleged to have founded Troy. That dating would be consistent with the dating of the divine allotments and the subsequent corruption of the Atlantians. What is more, it would accommodate Athens' prehistoric kings, Cecrops, Erechtheus, Erichthonius, Erysichthon, et al., who were enumerated in connection with the war against Atlantis (*Critias* 110A ff.), without necessitating the auxiliary hypothesis advanced by some scholars that all but Theseus were actual prehistoric kings, homonymous with a later, better known set ruling just before the Trojan War. As it happens, even the Cecrops and Erechtheus who ruled before the Trojan War were pictured by the Athenians themselves as part snake, part man (Cecrops) and all snake (Erechtheus). This strongly suggests that like the earliest royal genealogies of, say, Babylon and

Denmark, the Greek kings of prehistory are but astral and planetary deities in human pseudohistorical guise.

Ostensibly, then, there remains but one more obstacle to fitting the myth of Atlantis into the time-honored scheme of successive world ages. This obstacle is implicit in Plato's statement (*Critias* 119E) that the descendants of Poseidon's eldest son periodically engage "without weapons of iron" in a ritual hunt for a sacrificial bull. For if this means that iron was already available, the ritual hunt (which started with Poseidon's immediate successors) must be located in the Iron Age. That, however, conflicts with Plato's stress on the initial Golden-Age virtue of these successors and, fatally, with the fact that *ours* is the Iron Age. Hence it is best to take the phrase "without weapons of iron" to mean not that iron, though available, was forbidden, but that it was not yet available. This would place the hunting ritual in the Copper Age.

If, then, the Atlantis story is a Platonic variation on the familiar theme of successive world ages (and to that extent indebted to Solon's Egyptian informants), how are we to envisage the basic stages of the story? In the first place, the establishment of the cosmic frame, which forms the subject of the *Timaeus,* clearly belongs at the beginning of the Golden Age, the time when the Milky Way overarched the sky between Gemini (then the sun's place at vernal equinox) and Sagittarius (then the sun's place at autumnal equinox). These celestial conditions held from approximately 6000 B.C. to 4000 B.C.. The next phase is marked mythologically by the Olympians' conspiracy under Zeus against the Titans under Kronos, the defeat of the Titans, the relegation of Kronos, and the distribution of heaven and earth among the gods. This is the Silver Age, astronomically equivalent to the period (from roughly 4000 B.C. through 2500 B.C.) when at vernal equinox the sun was in Taurus, at summer solstice in Leo, at autumnal equinox in *Scorpius* and at winter solstice in Aquarius. This, too, is the time when "Poseidon" adopts Atlantis, Athena and Hephaestus Athens, and Athena alone Saïs (*Timaeus* 23E, *Critias* 109B). And, as we now know, it is the time when the zodiac was invented.

Nothing in the Atlantis story corresponds to the cannibalistic outrage (Lykaon's dish) which according to the mythological tradition closed the Silver Age and, astronomically speaking (about 2200 B.C.), marked the sun's shift of position at vernal equinox from Taurus to Pleiades. Yet this would seem to be the break so strongly marked by Plato's phrase, "But when the portion of divinity within them [the ten hereditary rulers of Atlantis] grew fainter . . . [i.e., when the dilution of their divine stock entailed their corruption]" (*Critias* 121A). Shortly thereafter the *Critias* breaks off. It seems safe, however, to infer that this corruption expressed itself, among other ways, in the unsuccessful attempt to conquer the Mediterranean world (*Timaeus* 25C), and that the Atlantid leaders' failure to learn the lesson of this defeat (where the Athenians figured as instruments of divine vengeance) caused Zeus to send (or authorize) the flood that indiscriminately destroyed both the entire island and the presumably occupying Athenian forces as well. Astronomically speaking, the flood would seem to have coincided with the sun's shifting (about 1300 B.C.) of its place at vernal equinox from Pleiades to Perseus.

Yet how does Plato's repeated statement that no fewer than 9,000 "years" have lapsed since the events of the myth square with the fact that on the astronomical grounds we have described, no more than 700 years can have lapsed since the Trojan War, no more than 1,630 years since the end of the Silver Age, and no more than 3,400 years since the end of the Golden Age? The difficulty is real though not fatal. One need not appeal either to the manifestly unfinished state of the *Critias* or to the Egyptians' familiar practice, long ago cited by Martin, of adding unequal yet consonantally homophonic time-units like the "year" ranging anywhere from a month to a solar year and beyond. Suffice it to recall that even in Greek and Latin the root meaning of "year" is merely (the quantitatively unspecific) "cycle." Thus, for example, Servius (on Vergil, *Aeneid* I 269 = Cicero, *Hortensius* fg. 35 Mueller = Aristotle, *Protrepticus* fg. 19 Ross) expressly distinguishes between "years" of 30 days, of 12 30-day months, and (as in the case of Aristotle's Great Year in the *Protrepticus*) of 12,954 solar years. Now the Atlantis myth itself combines with the Egyptian priests' own allusion to the doctrine of world-ages (*Timaeus* 22C ff.; *Critias* 112A) to argue that these 9,000 "years" cannot possibly be taken at face value. If, however, they are to be taken as so many 30-day months, they denote a period no longer than 750 schematic years. Which period, if added to 570 B.C., the putative date of Solon's Egyptian visit, brings us to 1320 B.C., i.e., to the very time-frame already inferred on other grounds as that of the Atlantid flood.

What then is the astronomical import of the war between Atlantis and Athens? If it is indeed astronomical, it can only refer to one of the two major phenomena which occur in the requisite eastward direction of Atlantis's attack upon the Mediterranean world, viz., to the planets' sidereal motion or to the gradual eastward shift of the constellations marking the sun's position on the equinoctial and solstitial days relative to a fixed and ancient horizon marker. The third century A.D. Neoplatonist Amelius opted for the planets, and indeed seems to have been the only one ever to have opted for an astronomical reading of the Atlantis myth, partial or entire. Amelius's preference, however, fails to account for the drowning of the innocent Athenians along with the corrupt Atlantians and fails, moreover, to provide a plausible tie-in with the doctrine of world ages expressly alluded to by the Egyptian priests.

In order to appreciate the other possibility, here being proposed as a solution to the problem, it is necessary to visualize the horizon-phenomena associated with the shift of the constellations at issue, and to see the eastern and western horizon phenomena as parts of a single process. At vernal equinox, the old constellation marking the sun's position "drowns" or "is drowned" by moving eastward beneath the horizon, while the new constellation replacing it descends from the sky on an eastward slant to take its position on the horizon. Yet the old horizon marker has to be shifted *westward* in order to accord with the new facts. (This was shown by the pair of illustrations in figure 1.) Diametrically across the sky, at the marker for autumnal equinox, the old constellation appears to be climbing eastward into the sky, while behind it a new constellation climbs eastward out of the "waters" onto the horizon. Here again, the old horizon marker has to be shifted westward to accord with the new facts.

It is worth noting in advance that, given the eastward direction of the Atlantid attack, there is nothing in the "correct," i.e., equatorial-ecliptical, construal of the phenomena at issue which would fit that direction. For if so construed, the only real motion of the equator along the ecliptic would be a westward one—which is contrary to the direction of the Atlantid attack. It is immaterial, therefore, whether or not Solon's Egyptian informants may be presumed familiar with the equator-ecliptic system (which, of course, Plato himself uses in the *Timaeus* in the course of describing the order of the universe). What alone matters here is the fact that an astronomical reading of Plato's Atlantis myth (other than Amelius's planetary one) is impossible except in terms of horizon-phenomena pure and simple.

What then is the most plausible correlation of the horizon phenomena with the several parts of the Atlantis myth? The mighty eastward push of the Atlantid naval forces best corresponds to the eastward advance from "water" onto "dry ground" of the new constellation marking autumnal equinox. Correspondingly, the Athenians' success in halting the Atlantid advance seems equivalent to their wresting the fiduciary horizon point from the Atlantians, who were hubristically seeking to carry it off in triumph in an eastward direction. The Athenians in effect wrest their marker free from the "drowning" old constellation and shift it to the west, i.e., contrary to the sense of the Atlantid advance. Mythologically, the old constellation "drowning" becomes its "being drowned" in punishment for the classical crime of presumption (*hubris*).

Yet even as they succeed in wresting the fiduciary horizon points from the enemy's suicidal eastward drive, the Athenians are themselves partially drowned in the process. On the island of Atlantis they lose to the flood their whole occupying army and navy. At home, in Attica, they lose most of their arable land to torrential rains. Their role of heroic self-sacrifice in effect anticipates that of the Greeks at Thermopylae. For there, too, a Greek sacrifice is redeemed by the certainty that it is sanctioned by the divinely established order of the universe. A similar point is made in marble on the frieze of the Nike temple on the Athenian acropolis. The Greek victories over Amazons and Persians are there portrayed as literally occurring under the eyes of the assembled gods, the implication being that the victory of the Greek cause, however costly, is the divinely sanctioned triumph of order and moderation over presumption and chaos.

Consistent, at least in principle, with an astronomical reading of the Atlantis myth is the fact that the myth is narrated to Solon in a place—Saïs in Lower Egypt—which along with the town and district of Canopus at the northwestern tip of the Nile Delta forms an integral part of an Egyptian system that correlated terrestrial and celestial locations. Following a Mesopotamian analog, the Egyptians construed their watery lifeline, the Nile, as a projection of specific segments of the sky, correlative and, at certain points, homonymous with celestial objects. Thus the references in the Babylonian creation epic to Eridu and Esagil are known to refer not, or not only, to the geographical places so named but also to stars, viz., Canopus, the bright star near the southern end of the ecliptical pole, and also Pegasus Square, a quartet of stars (α, β, and γ Pegasi and α Andromedae) describing a square whose heliacal rising marked (from 6000 B.C. through 4000 B.C.) winter solstice (figure 4). Similarly, in the (admittedly late) Egyptian temple of Dendera, the Upper Nile is taken as a projection of the zodiacal belt. To understand the projection, imagine the zodiac as inscribed on a circular band of paper. Cut the band at its southernmost point. The zodiac is now a long strip of paper with zodiacal north in its middle and zodiacal south at both of its ends (figure 5). Perhaps it is no longer quite so curious that zodiacal south coincides at once, in the Egyptian system, with the southern *and* the northern end of the Upper Nile. The principle of celestial geography involved undoubtedly antedates its application to the zodiac. Elsewhere, the same principle is applied to the Lower Nile, necessarily in much greater compression. Here the town and district of Canopus at the northwestern tip of the Delta and an island in the extreme south of Upper Egypt are both taken as terrestrial equivalents of Canopus, while the town and district of Saïs, southeast of Canopus, is "hieratically [in priestly parlance]" (so Proclus) taken to represent the north pole.

Figure 4
A Mesopotamian representation of Pegasus Square, a quartet of stars comprising α, β, and γ Pegasi and α Andromedae. The square itself appears at the upper right of the illustration; the beings to either side of it are probably astral deities. Beneath the square, the sacrifice of an animal is portrayed. The entire image is taken from the surface of a seal that was used to press the symbols into clay. Note the resemblance of Pegasus Square to the arable rectangle (see figure 2) created by Poseidon's descendants on Atlantis. In support of this analogy is the fact that the Babylonians referred to the Square as the "primal field."

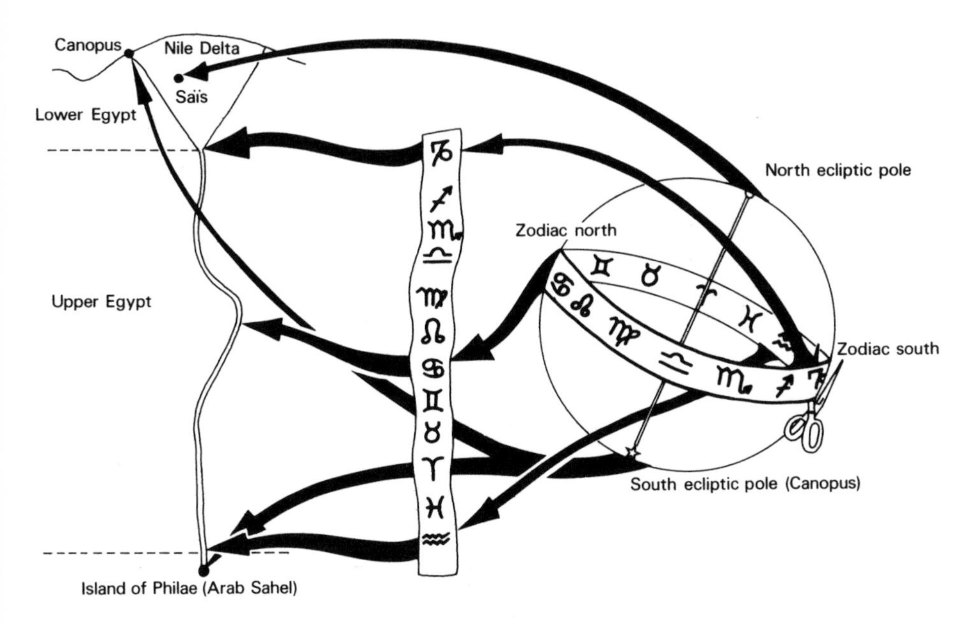

Figure 5
A system by which the ancient Egyptians correlated locations in the sky with locations on earth. In the rightmost part of the illustration, the band of the zodiacal constellations is shown, together with its central axis, terminating in the north and south poles of the ecliptic. The south pole is marked by the star Canopus. The modern constellations of the zodiac have been placed on the celestial band, solely to demonstrate the nature of the projection system at issue; those constellations do not represent the way the Egyptians would have construed the sky, and in any case they are shown in their modern positions, with Capricorn, for example, at the southernmost point of the ecliptic—the point of winter solstice. Now the nature of the celestial-terrestrial mapping shown in this figure was elucidated by Georges Daressy, who analyzed an Egyptian zodiac—specifically a zodiac represented along the architraves of several Egyptian temples as a sequence of animals. Below such sequences is a geographically accurate representation of the Nile, which alone would suggest a one-to-one mapping of sky to earth. However, Daressy also knew that particular townships along the Nile had as their emblems the same or similar animals. Note that the zodiac in the illustration is cut at its southermost point, and that the resulting linear sequence of constellations is mapped only onto the portion of the Nile south of the Nile Delta—the portion in Upper Egypt. If the Egyptians mapped the zodiac onto the Lower Nile, the details of that mapping remain to be discovered. Note also that several further aspects of the Egyptian association of the sky with the earth are shown in the illustration: The south pole of the ecliptic is mapped onto the island of Sahel, but it is also mapped onto the earthly town of Canopus, at the westernmost point of the Nile Delta in Lower Egypt. The north pole of the ecliptic is mapped onto Saïs in the Nile Delta.

Now it is at Saïs, the terrestrial representative of the north pole, that the Atlantis tale originates. And it is (as we shall now see) to Canopus, visible as far north as the 36th parallel, and to the southern circumpolar area of the sky that the story seems actually to refer.

Much favors the working hypothesis that Poseidon's artificial island (within the larger island of Atlantis) is in fact to be sought in the vicinity of the south pole of the ecliptic. In the first place, the ecliptic south pole was thought to be marked by the star Canopus and to be immune to the passage of time; it is the "island" to which Kronos/Saturn was banished upon his downfall at the end of the Golden Age—the island upon which Kronos sleeps while time stands still. And indeed, being near the south ecliptic pole, Canopus is in truth more or less immune to the passage of time—astronomical time. It is more or less immune, that is, to the drift of stars and constellations that entails the succession of world ages. Consider now Plato's express statement (*Critias* 119E) that near the center of the artificial island in Atlantis was a metal pillar upon which Poseidon's ten immediate descendants had inscribed his laws, and where they and their successors periodically assembled. It was over this pillar that the throat of a sacrificial bull was cut at such assemblies—the more appropriately so if the pillar was (or represented) the cosmic axis at its southern pole and the offering, the bull, represented the constellation Taurus. For the Silver Age, to which we have tentatively assigned this portion of the Atlantis tale, was the time when the rise of the constellation Taurus marked the arrival of the vernal equinox. That the pillar is indeed the cosmic axis is independently suggested by Homer's resting the axis of the universe on Atlas's island, presumably "at the navel of the sea." (Recall that Poseidon gave over the rule of his artificial island to his first-born, Atlas.)

Numerous further features of the tale of Atlantis also appear to reproduce specific features of the southern circumpolar sky. Consider that what the Babylonians termed "the confluence of rivers," the place where Enki-Ea, the Babylonian Kronos, dwells, is the joining in this portion of the sky of the two arms of the Milky Way: one appearing to stream down past Orion, Sirius, and Canopus; the other funneling down through the gap guarded by Scorpio and Sagittarius, and then past the Altar (Ara), usually depicted as flaming or smoking, where the Olympians are reported by Eratosthenes to have sworn their oath of allegiance to Zeus before setting out to battle the Titans and depose Kronos—the events that ended the Golden Age. There is, moreover, the poop of a ship, Argo, elaborately equipped with sails and oars (Canopus being the rudder), which seems to be sailing along an arm of the Galaxy. (Consult the sky-map of figure 6.) Now the tale of Atlantis expressly mentions (*Critias* 116E) an altar in the vicinity—presumably in front—of the temple of Poseidon, the temple in which the metal pillar stood. Upon this altar the flesh of the sacrificial bull was burned. What are we to make of this? Note that Ara is diametrically across from Canopus in the southern sky. It looks, therefore, as if we have in Ara and Canopus the celestial prototype of the altar and pillar mentioned in the myth. The myth of Atlantis also explains that, to get to and from the innermost island of Poseidon, a ship could pass beneath fortified bridges overarching the canal constructed by Atlas's descendants. This detail, too, has a celestial prototype: the poop of the Argo, northbound in the Milky Way. For the overarching bridges would conceal from view the center and bow portions of a ship that had begun a passage beneath such a bridge.

Yet if the entire section of the southern sky from the south pole to about latitude $-50°$ is to be equated with the central island of the Atlantis myth, then the concentric bands of water and earth with which Poseidon and his successors ring this island must be taken to represent celestial latitudes reaching at least as far as the central area of the sky, which we describe in terms of equator and ecliptic, the Babylonians as the way of Anu, and the early Egyptians as a belt of stars, led off by Sirius. Happily, that very supposition is entirely consistent with Plato's statement that on the second of the circular bands of land (counting outward from the central island) Atlas's descendants laid out an "equestrian race-track" flanked on either side by "barracks for the greater part of the spearmen." (*Critias* 117C). After all, the central area of the sky is the realm of the planets, and the association of the differential planetary revolutions with races and racetracks is standard throughout antiquity (being reported in connection even with the Roman Circus Maximus). Moreover, the central area of the sky is the realm of the constellations composing the zodiac. And those constellations are normally pictured as "houses." Furthermore, other subdivisions of this central area of the sky, both temporal and spatial, are pictured as people or as "warriors."

Figure 6
The southern sky as depicted by Jan Hevelius in an atlas of the heavens published in 1687. The Milky Way (though not explicitly shown) streams more or less vertically in this illustration. It flows between Scorpio and Sagittarius at the top of the circular projection, and then past Ara, an altar here depicted as aflame. Toward the bottom of the illustration, it flows behind the head of Canis Major and then passes Orion near the perimeter of the circle. The poop of a ship (the constellation Argo) appears to be afloat in this milky river; the star α Carinae, or Canopus, second brightest in the sky, is the most prominent of the two stars defining the rudder of the ship.

If, then, the "sea" linked by canal to the central island represents the sky at large, the Galaxy represents that very canal, and the concentric bands represent circular areas of the sky, what does the enormous irrigated plain represent in celestial terms? (Recall that the rulers of Atlantis made an irrigated rectangle out of the arable portion of Atlantis island.) Pegasus Square leaps to mind as the most likely candidate. For the Babylonians regarded Pegasus Square as the primordial field; and they made it "leader" (in a Babylonian star catalogue ultimately reflecting the period of around 1100 B.C.) of the stars in the way of Ea (roughly the lower third of the sky).

One final category of evidence ought now to be weighed: namely, what significance can we attribute to the numerical values embodied in the dimensions of the island-within-an-island designed by Poseidon?

Figure 7 shows the proportions of Poseidon's island. One conclusion is especially striking: What Poseidon had instructed his successors to do, and what those successors had inscribed on a metal pillar, viz., "to give equal honor both to the even and the odd" (*Critias* 119C), Poseidon himself had done first; for the numbers 2 and 3, with their sum (5) and product (6), appear repeatedly in the construction. Consider that Poseidon's construction extends to the fifth band in the illustration, i.e., to the band with the outer diameter of 27 stades. (Poseidon's successors added the sixth band.) The original construction, designed by Poseidon, thus comprises six (i.e., 2×3) parts: a central island with a diameter of five (i.e., $2 + 3$) stades; and then five concentric bands, three of them destined to be filled with water and the remaining two, intercalated in between, to be annular bands of land. The three bands of water have an aggregate width of six (i.e., 2×3) stades; the two bands of land have an aggregate width of five (i.e., $2 + 3$). To top it off, the area of the outermost band of water is 216 square stades (i.e., $2^3 \times 3^3$). The reason for all this undoubtedly has to do with the Pythagorean-Platonic practice of classifying certain numbers by means of a colorful symbolic shorthand unfairly dismissed as "number mysticism." In this system, two and all numbers divisible by two are termed "female" and all odd numbers (starting with three) are termed "male"; their sum (five) and product (six) are consequently spoken of as being "marriage" numbers. It seems, therefore, that much as a Renaissance prince might adorn the

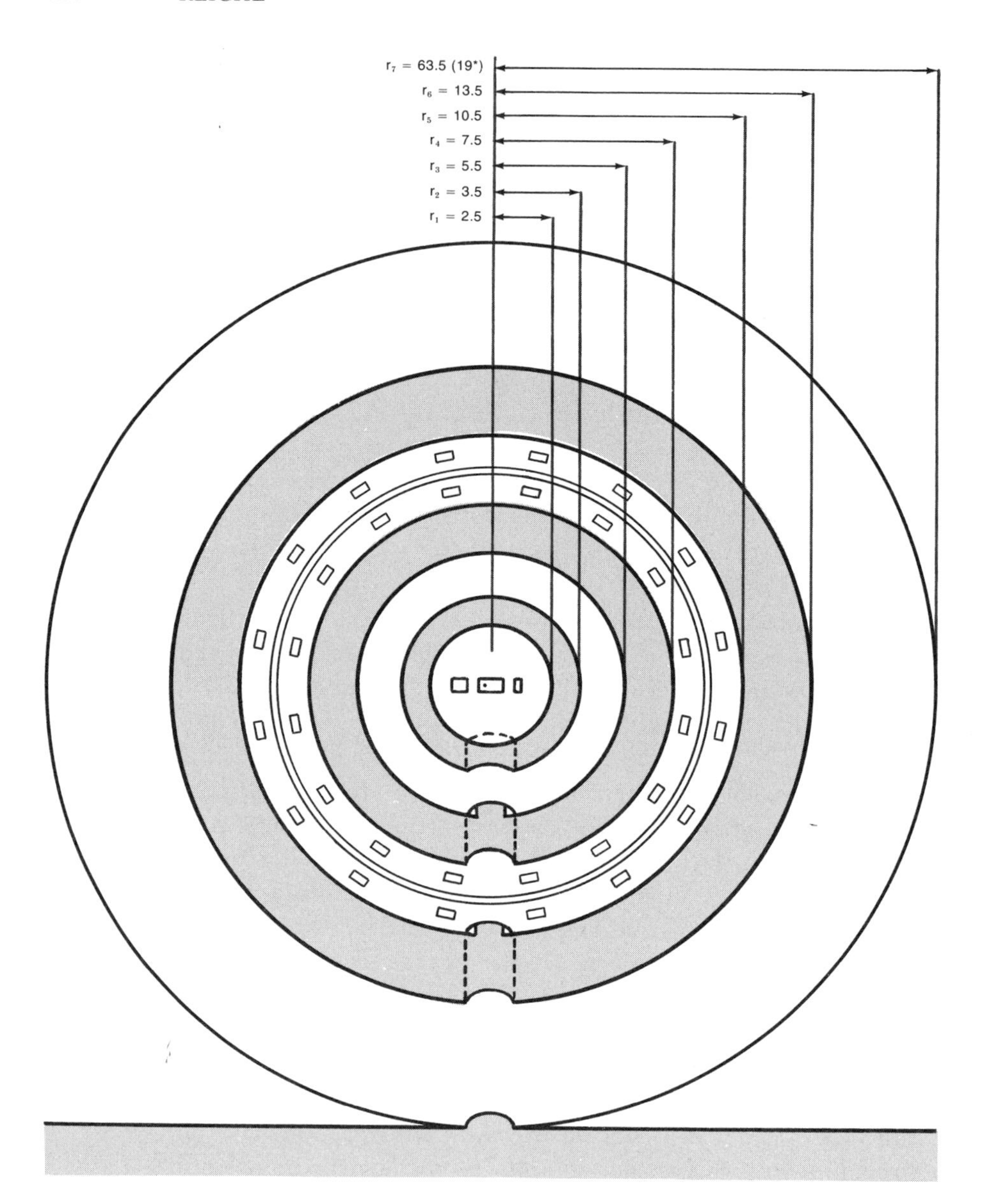

r7 = 63.5 (19*)
r6 = 13.5
r5 = 10.5
r4 = 7.5
r3 = 5.5
r2 = 3.5
r1 = 2.5

Figure 7
Poseidon's constructions on the island of Atlantis, as specified by Plato in the *Timaeus* and the *Critias*. Poseidon is reported to have created a series of rings, two of land and three of water, around an innermost island. Beyond that, Plato gives few details. Still, we are told of an altar on the innermost island, and also of a Temple of Kleito and Poseidon (Kleito is the mortal woman with whom he mated). The temple is said to be "at the center." Moreover, we are told that a metal pillar with Poseidon's laws inscribed thereon was "near the center of the island in the temple of Poseidon." The design shown here for the innermost island is patterned in large part on the Aphaia temple of Aegina, an island near Athens. The temple is indeed at the center; the pillar is within, though not quite at the precise focus of Poseidon's constructions (a statue of Poseidon is presumed to occupy that place). To the east of the temple is the altar; to the west is a palace. Additional features of Atlantis, added by Poseidon's descendants, include an "equestrian ractrack" and "barracks for the greater part of the spearmen" on the second annular band of land. There is, in addition, a canal linking the central island with the sea, and passing beneath covered arches, shown here by dashed lines. The radii defining the concentric bands of Poseidon's work are given at the top right to the illustration; they are the values set by Poseidon himself, and are expressed in stades.

* The outermost band of earth was constructed not by Poseidon but by his descendants. It is shown here as if its radius were 19 stades, which is the appropriate value if, as explained in the text, each radius ought to be larger than the preceding radius by a factor of $\sqrt{2}$. Poseidon's descendants, however, gave it a radius of 63.5—more than three times too large for the scheme established by the builder of the island.

boudoir of his mistress with the motif of his and her initials intertwined, so does Poseidon, in Atlantis, surround his marriage bower (recall that he mated with a mortal woman on Atlantis) with the numerate equivalents of marriage.

But this is not all. Poseidon's choice of the central island's radius (2.5 stades) is such—and this appears to have gone unnoticed in the scholarly literature so far—that of the six radii defining the annular bands of earth and water, the second, fourth, and fifth are practically equal to the first, third, and fourth as multiplied by $\sqrt{2}$. The third radius is rather too large to be $\sqrt{2}$ times the second (5.5. instead of 5); but its excess is corrected by the corresponding shortfall of the sixth (13.5 instead of 14.8). Apparently, Plato had two constraints to meet—that of having Poseidon equally honor the even and the odd, and that, it seems, of creating the progression we are here considering. Thus it appears valid to note that each of the radii is greater than the preceding radius by a factor approaching

$\sqrt{2}$; that is to say, the progression of radii can be expressed very nearly in the form $r_{n+1} = \sqrt{2}\, r_n$. Accordingly, if all six radii are graphed as in figure 8, it can readily be seen that each radius turns out to be the side of a square whose diagonal is the next larger radius. This is a point that was of considerable importance to the early Pythagorean mathematicians. For the Pythagoreans believed, as an extrapolation of data derived from music and astronomy, that progressions of integers (or musical intervals composed of integers) recurred throughout the universe: in the successive realms of number, figure, solid, vibrating string, and revolving planet. Doubtless, therefore, they were horrified to find, upon constructing a square of unit size, that the diagonal of that square has an irrational value, and thus is "incommensurable," to use their term, with the rational—indeed, integral—measure of the side. Doubtless, too, they were relieved to find that commensurability could be restored at the next higher dimension—that of plane geometry—by taking the diagonal of the square to be the side of a larger square. For the area of the larger square is once again commensurable—it has a rational value. In short, therefore, an apparent incommensurability at the level of arithmetic is reconciled at the level of plane geometry, and nothing has occurred to invalidate the Pythagorean axiom of universal, i.e., cosmos-wide, commensurability.

In his old age Plato is known to have postulated the extension of this principle (of restoring commensurability by moving to successively higher dimensions) to include solid geometry and kinematics, both musical and astronomical (*Epinomis* 991D-992A). In the case at hand, Poseidon performs the transition from the first to the second dimension no less than six times in succession. The principle is important in view of "Poseidon's" manifest identity (see above) with the Engineering God Kronos/Saturn—the same, presumably, who in the *Timaeus* had functioned as engineer of the entire astronomical cosmos. In effect, Poseidon/Kronos is proclaiming, from his present "island" abode at or near the celestial south pole, that same universal commensurability which, in a previous "Golden" world era, he had embodied in the astronomical universe at large. For we have already seen that the rings surrounding the central island of Atlantis can be understood as bands of the sky reaching all the way to the middle, equatorial-zodiacal realm.

Visually, the spiral we have generated in figure 8 by connect-

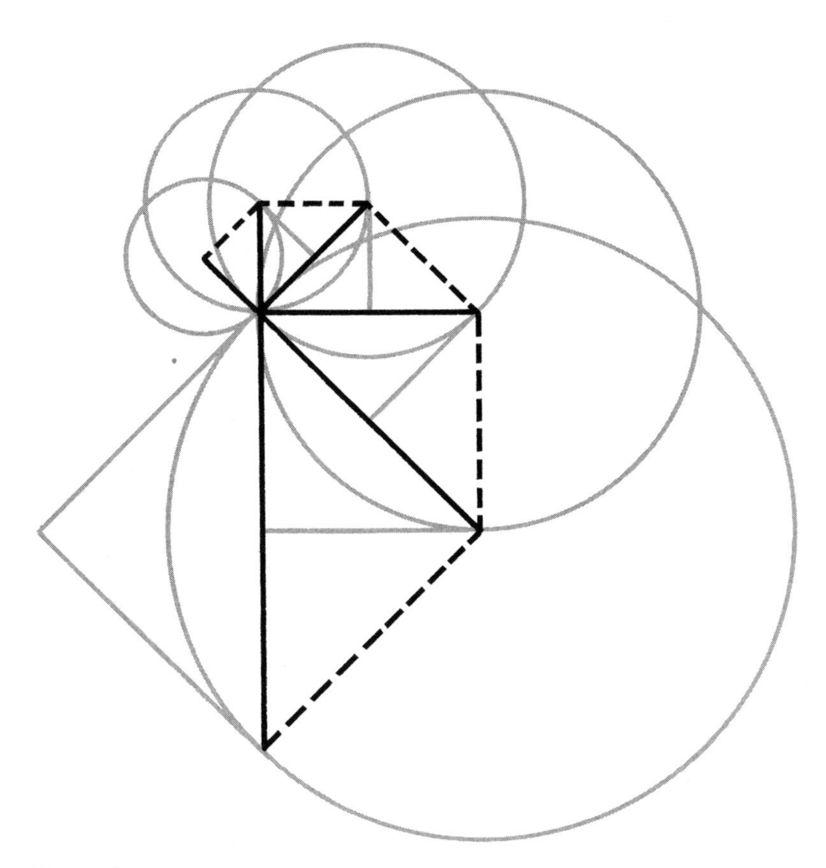

Figure 8
"Poseidon's conch": a geometrical construction made by the author to demonstrate a possible significance in the radii chosen by Poseidon for the concentric bands of earth and water with which he ringed his sanctuary on Atlantis. In the construction, each of the six line segments shown in black is first made the side of a square, and then the diagonal of a larger square. The sequence of lines, if one begins with 2.5, turns out to be 2.5, 3.5, 5.0, 7.0, 9.9, and 14.0. These are close to the values for the radii of Poseidon's island as shown in figure 7. (If one multiplies each of Poseidon's radii by $\sqrt{2}$ and compares the result with the value of the next larger radius, the fit is even closer.) For the significance of converting the side of a square to the diagonal of a larger square, consult the text. The outer sides of the consecutive squares are shown here as dashed lines. Note that they form the stylized outline of a conch-shell—an appropriate symbol for Poseidon, who dwelt in the sea and whose entourage included Capricorn, blower of the conch-shell to halt the Deucalian flood.

ing the successive radii of Poseidon's central island resembles a marine conch-shell. This is a happy circumstance, for both generally and specifically, the conch is ideally suited to serve as Poseidon's personal emblem. It is generally suitable because the conch is traditionally blown by members of Poseidon's marine entourage in the briny deep, which in this case is the area near the celestial south pole to which, like the departed souls, one descends by way of a portal at the southernmost point of the zodiac. Specifically it is suitable because the conch was reportedly invented by Capricorn, the watery figure associated with winter solstice at the southernmost point of the zodiac. The blowing of the conch-shell by Capricorn, incidentally, signaled the end of the Deucalian flood. Have we, then, come upon the tell-tale symbol of "Poseidon's" power over the "waters of the deep" which engulfed Atlantis? After all, the power to generate earthquakes and tidal waves is traditionally his rather than Zeus's. Consider, too, the directives issued by Poseidon to his ten sons and inscribed by them on a metal pillar, yet evidently disregarded by his remote descendants at the stage of their corruption. Consider, in particular, that the oath taken by Poseidon's sons invoked "mighty curses upon them that disobeyed" (*Critias* 119E). It is obvious from the width of the land-belt added by Poseidon's descendants (50 stades) that the "law" of Poseidon's conch has been broken. What should have been a radius of 19 stades turns out to be 63.5 stades—more than three times too large.

This article has offered, in briefest outline form, the case for considering the layout of Plato's Atlantis as an embellished version of what in original intention was a map of the sky, and the battle of Athens against Atlantis as but a pseudohistorical description of one aspect of a continuous process that becomes critical about every 2,000 years—and not a memory of lost islands or continents. There remains, however, a rather general question that may trouble contemporary readers: If our account, or something like it, is indeed the underlying meaning of Plato's myth of Atlantis, why did not Plato state that meaning plainly and overtly? Why should he have chosen to mystify us (if that is the right term) both as to the ultimate referent and as to the logic informing his own numerical elaboration of the myth?

The answer is surely that Plato was heir to the ancient and

widespread conviction that the sky is the place of the (astral) deities and at once the origin and ultimate destiny of our souls. Astronomical knowledge is therefore at once scientific and sacred. Indeed, the very act of formulating the numbers and figures that govern the behavior of astronomical bodies is a method by which to enhance our intellectual and moral qualifications for posthumous reassimilation to the starry realm from which we came. (In nonphilosophical circles, technical astronomical knowledge was, of course, additionally credited with putative power over the celestial bodies themselves, hence for that reason alone treated as classified information. Philosophical materialists like Lucretius actually insisted that this magical notion of astronomical knowledge remained latent in Platonism.) Thus, the inherent difficulty of astronomical observation, modeling, and computation combines with the belief in the moral import of this purely "theoretic" activity, and out of this fusion comes a style of scientific communication verging on the esoteric. Plato's motive, then, is not the (to us) familiar one of safeguarding one's intellectual property prior to (or in the absence of) patents and copyrights. Rather, it is a quasi-religious respect for the dignity and importance of the subject matter. The proper format for handing it on is not indiscriminate (i.e., written) dissemination but oral instruction of intellectually and morally qualified pupils only.

Surely this explains the prolonged resistance by the ancients to the translation of cosmological and astronomical information into written form, and specifically into mathematical notation. That the traditional, oral form of transmission was not due to the unavailability of a written alternative is shown by Caesar's description of the priestly discipline as practiced in Gaul (*The Gallic War* VI 14): "It is said [reports Caesar] that the young apprentice priests have to memorize a large number of verses and that some of them spend as long as twenty years in training. And they [the Druids] *think it irreverent to commit these verses to writing* [emphasis added], although they use Greek letters for almost all other public and private affairs." Somewhat superficially, Caesar first traces this belief to a mere penchant for secrecy. Later on, though, he ascribes it, more justly, to the fear lest recourse to writing dull the powers of memory. Just so in the *Phaedrus,* Plato has the Egyptian god Amon object to Mercury's invention of writing on the grounds that it will weaken men's memory and foster the illusion that the dead letter of mere memory-aids can produce a knowledge equivalent to that born only from living exchange with a master.

Caesar then continues: "A central tenet of [the Druids'] teaching is the immortality and posthumous transmigration of souls, a doctrine calculated to inspire death-defying courage. To this end they discourse on, and transmit to the young, many matters concerning stars, constellations, and planets [the Latin term *sidera* covers all three], including their motions, the size of the earth and its several regions, the nature of the universe, and the power and influence of the imperishable [because astral] deities."

In conclusion, some more general comments on this preference of oral and mythological to written and conceptual expression in matters cosmological and astronomical may be appropriate.

Originally, no doubt, the use of mythological expression was not a matter of preference but of the absence of alternatives. The very perception of astronomical phenomena appears to have been configurational rather than positional. Thus, a planet was (and in some quarters of the world still is) taken not, or not solely, as the luminous point here and now visible but as the (almost) closed geometric figure which it regularly traces out in the zodiac over a period of time as short as a month or as long as several years, depending on the planet involved. And the form of expression best suited to such configurational perception is, of course, the narrative form, a tale ("myth" in Greek). In a purely oral, memory-based culture, there would in any case exist no real alternative to the narrative format. Add to this the primeval conviction that naming, enumerating, describing something constitutes a form of incipient action, of getting involved with that something; that the very terms of referring come charged with a power, for good or ill, over the referent. In astronomy, this conviction demonstrably survived well into the so-called scientific era. Thus, long after the Athenians had grasped the mechanism and mathematics governing eclipses, they maintained a special priesthood charged with interpreting their religious import. It follows as a corollary that the terms of referring, given their potential hold over the referent, must be treated with respect or, if not as outright taboo, at least as "classified information"—in the interests at once of the astronomical referent and of the human referrer. The astronomical referents themselves, of course, are conceived as divine; mathematical predictability and religious significance, indeed "personality," being deemed perfectly compatible. In effect, mathematical number and law are conceived as providing statistical rather than fully deterministic constraints upon the deities composing the astronomical universe and as, in that sense, predicting their observable behavior—without plumbing the Reasons for cosmic and human destiny. Yet even while the theoretical component of piety strives for ever greater accuracy in those predictions, its pragmatic component seeks to mute that very progress by retaining, if in an increasingly symbolic sense, the most ancient forms of address and description. In the case of the Doric temple, this priestly conservatism notoriously accounts for the duplication in marble of structural details obviously associated with an originally wooden structure. In the case of astronomy and cosmology, it accounts for the retention of pseudohistorical, pseudogeographic, and pseudozoological modes of expression long after purely numerate alternatives have become available. The quaint language in terms of which astrology describes the mutual relations of the planets is a case in point. Priestly conservatism also explains the treatment of the numerate alternatives as "classified information." And it explains, finally, the insistence that even these alternatives be memorized rather than committed to writing, long after writing had come into use.

In sum, the priestly experts in charge of archaic astronomy, and their spiritual descendants throughout the classical portion of recorded history, had to reconcile four partly conflicting tasks: that of conserving, that of improving, that of transmitting, and that of protecting the sacred astronomical-cosmological knowledge. They thus effectively combined the heterogeneous functions of communal memory, pure research, mouthpiece, and censorious custodian.

Between these several constraints, then—to adapt Gibbon's famous comment on the Nicene creed—the almost invisible and tremulous ball of astronomical-cosmological knowledge could be allowed securely to vibrate. The result: much of "mythology" as we know it—and Plato's cosmological myths, including, so it has here been argued, that of Atlantis.

Seen in this light, the special hybrid character of Plato's Atlantis myth comes into focus. It is now recognizable as a piece of sacred cosmology deliberately expressed in the pseudohistorical and pseudogeographic terms familiar from "mythic" language in its ancient capacity as a technical shorthand for astronomical systems. That Plato retains it to describe the stellar equivalent of what since Hipparchus we describe as the precession of the equinoctial points—and this at the very time that Greek astronomy took its first great strides—says much about the ultimate importance he attaches to the subject. That he derives this particular mythic embodiment from a culture as ancient and "catastrophe-proof" as that of Egypt is essential to his case for the periodic achievement, loss, and recovery of high culture as an integral part at once of the "life" of the cosmos and of man's moral progress. The translation of Egyptian into Greek deities and the demonstrable mistranslation (probably due to Solon) of at least one item—the "9,000 years"—are a small price to pay for that grand enterprise.

For further reading:

Basic works on the problem of oral composition are Milman Parry, "Studies in the Epic Technique of Oral Verse-Making: I. Homer and Homeric Style" and "II. The Homeric Language as the Language of an Oral Poetry," *Harvard Studies in Classical Philology* **41** 73–147 (1930), **42** 1–50 (1932); and Albert Lord, *The Singer of Tales,* Harvard University Press, 1960. Particularly helpful for an understanding of the social and psychological implications of oral transmission is Eric Havelock, *Preface to Plato,* Harvard University Press, 1963.

Apart from Gerald Hawkins's papers on Stonehenge (reprinted as appendices to his *Stonehenge Decoded* and *Beyond Stonehenge,* New York, 1973) and Alexander Thom's *Megalithic Sites in Britain* and *Megalithic Lunar Observatories* (Oxford University Press, 1967 and 1971 respectively), the best introduction to the field of archaeoastronomy is probably afforded by the little book of Rolf Müller, *Der Himmel über dem Menschen der Steinzeit* (Berlin, 1970), and by the special collective issue of the *Philosophical Transactions of the Royal Society of London* (volume 276, 1974) entirely devoted to a symposium on the place of astronomy in the ancient world.

Regarding Professor von Dechend's work, the most accessible publications to English-speaking readers are still Giorgio de Santillana and Hertha von Dechend, *Hamlet's Mill,* Gambit Press, Boston, 1969—with my review in *Classical Journal* October/November 1973, 81–83—and two large folders of mimeographed lectures delivered at MIT in the sixties in the course of several seminars on archaic cosmology. In German there is a great deal more: a chapter on the "Donnerkeil" in *Prismata: Naturwissenschaftsgeschichtliche Studien. Festschrift für Willy Hartner,* edited by Maeyama and Saltzer (Wiesbaden, 1977, pp. 95–118), and fifteen typed, bound volumes, averaging 120 pages in length, of her Frankfurt University lectures and seminars since 1970. The English translation and publication of this important research is now being undertaken.

On Plato's Atlantis myth, there is, of course, an immense literature that starts with antiquity. A good first survey up to the middle of the nineteenth century is afforded by Thomas Martin's *Etudes sur le Timée de Platon,* Paris, 1841, volume I, 257–333, and, for subsequent years, by Hans Herter, "Platons Atlantis," *Bonner Jahrbücher* **133**, 28–47 (1928), and by Jean Bidez, *Eos ou Platon et l'Orient,* Brussels, 1945, pp. 19–40. Advocacy of the Santorin hypothesis of Marinatos and Galanopoulos can be found in Mavor, *Voyage to Atlantis,* New York, 1969.

For Egyptian and Babylonian astronomy, consult G. van der Waerden, *Science Awakening,* volume II: *The Birth of Astronomy,* New York, 1974; and for Egyptian celestial geography, consult Daressy, "L'Egypte Céleste," *Bulletin de l'Institut Français d'Archéologie Orientale du Caire,* XII, 1–34 (1915).

For circular and spherical models in Greek astronomy, consult Dreyer, *A History of Astronomy from Thales to Kepler,* New York, 1953, pp. 53–122, and Walter Saltzer, *Theorien und Ansätze in der Griechischen Astronomie,* Collection des Travaux de l'Académie Internationale d'Histoire des Sciences No. 23, p. 65 (Wiesbaden, 1976).

For Great Years as periods equivalent to the least common multiple of various astronomical subperiods, as intervening between floods and/or fires, and as marking the repetition of some or all events, see Van der Waerden, "Das Grosse Jahr und die ewige Wiederkehr," *Hermes* 80 (1952) 129–55. For an exhaustive typology of these notions and an explanation of Aristotle's Great Year of 12,954 solar years in particular, see my forthcoming "Aristotle's Great Year of 12,954 Years (*Protrepticus* fg. 19): Explanation and Background." See also my "Gallehus Horns, Lunar Declination Cycle, and Ragnarok," in *Prismata* (Wiesbaden, 1977), esp. pp. 321–24.

For the archaic character of the rituals performed in Atlantis, see Hans Herter, "Das Königsritual der Atlantis," *Rheinisches Museum* **109** 236–59 (1966).

For the numbers incorporated by Poseidon into Atlantis, see the important exchange between Brumbaugh," "Note on the Numbers in Plato's *Critias,*" *Classical Philology* **43** 40–42 (1948), and Rosenmeyer, "The Numbers in Plato's *Critias*: A Reply," *Classical Philology* **44** 117–20 (1949). My own analysis is in preparation.

Finally, on the materialist charge that even Platonism, in the works of Plato and Aristotle, is still mythological, see my "Myth and Magic in Cosmological Polemics: Plato, Aristotle, Lucretius," *Rheinisches Museum* 114, esp. pp. 307–16 (1971).

Two papers in the recent *Atlantis: Fact or Fiction* (Indiana University Press, 1978) bear closely on the foregoing study. Luce (pp. 65; 76–8) sheds new light on the identity of Plato's narrator. Fredericks (pp. 86; 93) acknowledges the Near Eastern background of Greek mythology, but, unaware of the capacity of myth to function as vehicle for technical, astronomical information, dismisses this as Euhemerism.

THE STONEHENGE DECODER

On the following pages will be found two elements that can be cut out and assembled to show the year-round motion of the sun over Stonehenge. To assemble the device, cut out the strip on page 193 and tape it to a twelve-ounce beer or cola can. (If you prefer not to cut pages from books, you can try tracing or copying the elements of the decoder; unfortunately, all copying machines enlarge material slightly.) The can now represents the motion of the sun through the celestial sphere. In particular, that motion can be decomposed into two parts. The first of these, caused by the earth's turning on its axis once each day, is simulated by turning the can on its axis in a clockwise direction. The second, caused by the earth's revolution around the sun each year, moves the sun along the so-called ecliptic, the curved line on the paper strip, one complete cycle each year. To visualize the superposition of these motions, the reader is urged to imagine a roulette wheel spinning in one direction as the sun moves along its track in the other.

Now cut out the element on page 195, tape it to a piece of cardboard, and cut out the oval hole. Position the cola can inside the hole in such a way that the edge of the hole closely meets the wall of the can, and tilts in the direction shown on the decoder. When positioned properly, the uppermost point on the oval's edge will touch the strip of paper along the line marked "north circumpolar boundary," and the lowermost point on the edge will touch it along the line marked "southern horizon."

To find the motion of the sun over Stonehenge on any given day, simply choose a point on the ecliptic and follow its motion as the can is turned clockwise, always keeping the can properly tilted relative to the oval. At spring or fall equinox (corresponding to sun-images labeled "March" and "September"), the sun rises due east and sets due west. At summer solstice (sun labeled "June"), the monument reveals its solar alignment: on this day, when the sun is highest over the celestial equator, it rises over the Heel Stone and in alignment with the central axis of Stonehenge, as marked out by the Heel Stone and the central trilithon. At winter solstice, the sun (labeled "December") sets in alignment with the other end of the axis.

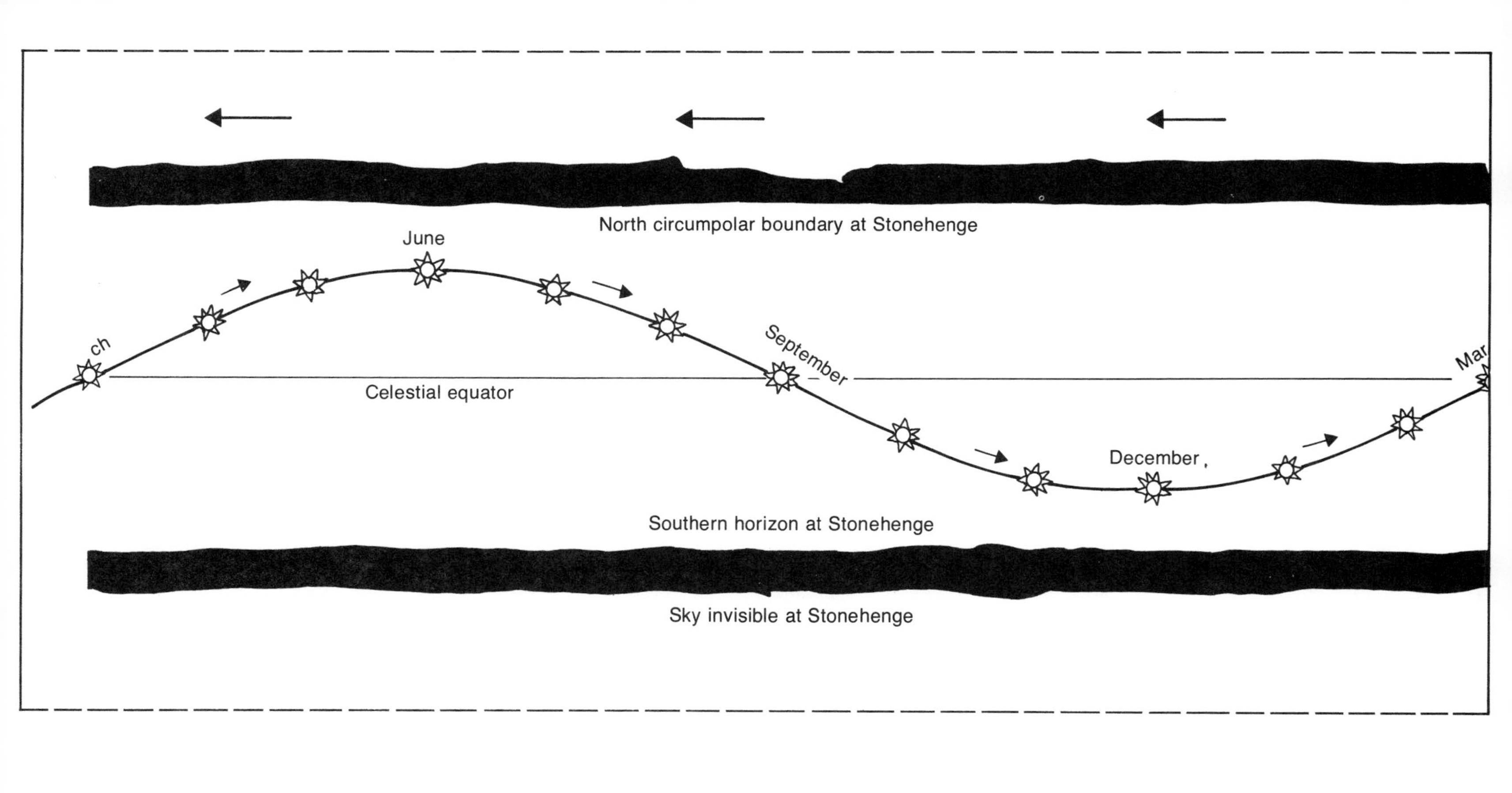
North circumpolar boundary at Stonehenge
June
ch
September
Mar
Celestial equator
December
Southern horizon at Stonehenge
Sky invisible at Stonehenge

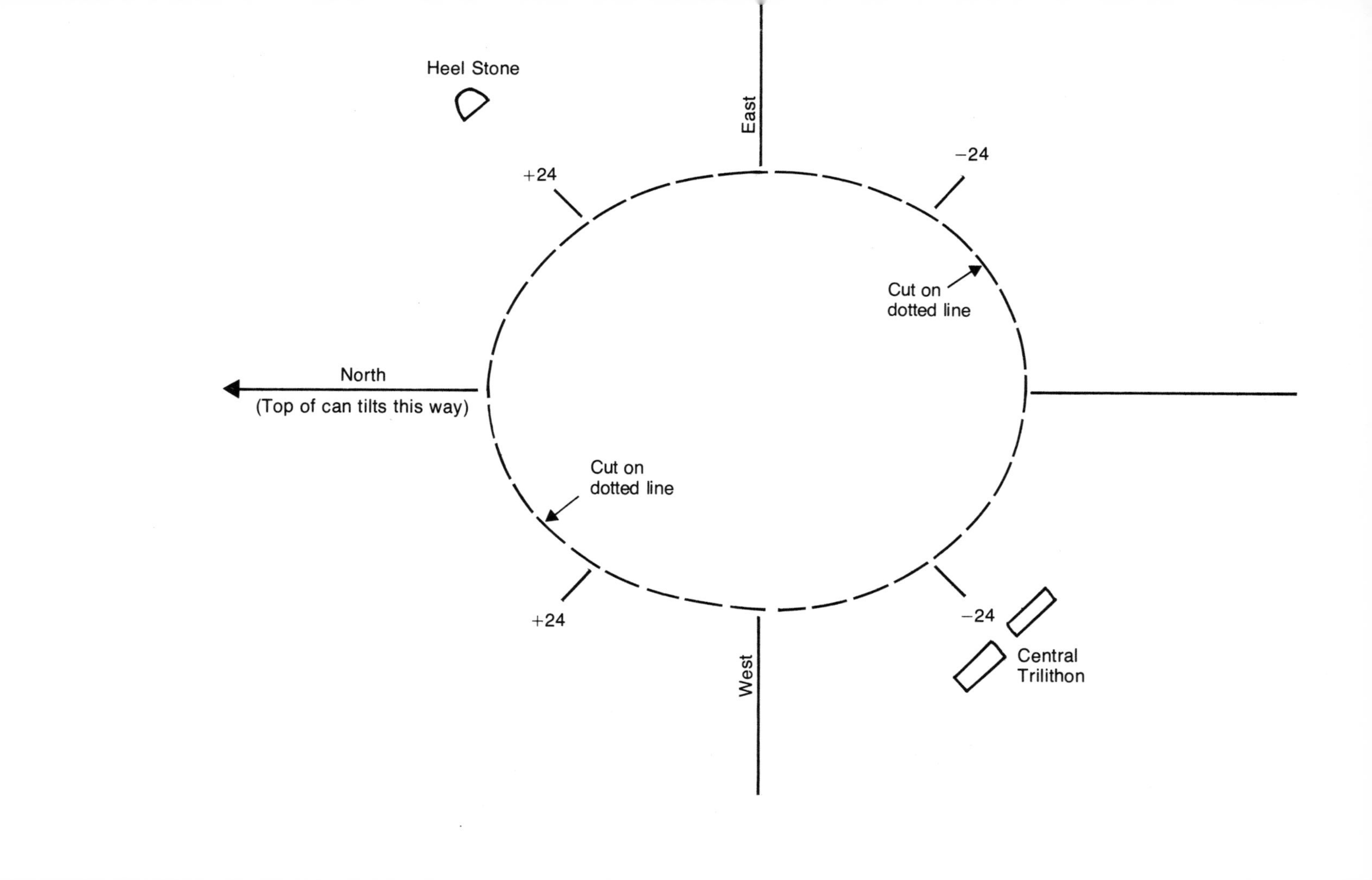
Heel Stone
East
+24
−24
Cut on
dotted line
North
(Top of can tilts this way)
Cut on
dotted line
+24
−24
West
Central
Trilithon

CONTRIBUTORS

Anthony F. Aveni is Professor of Astronomy at Colgate University. His recent activities have centered around research in the field of archaeoastronomy, primarily of ancient Mexico. He is editor of three books on archaeoastronomy and the author of numerous papers and articles on the subject.

John C. Brandt is Chief of the Laboratory for Astronomy and Solar Physics at the Goddard Space Flight Center of the National Aeronautics and Space Administration.

Kenneth Brecher is Associate Professor of Physics at MIT and a member of MIT's Center for Theoretical Physics and its Center for Space Research. His research interests are high-energy astrophysics, general relativity, and cosmology. Most recently, his work has centered on astronomical tests of the fundamental laws of physics.

John A. Eddy is a Senior Scientist on the staff of the High Altitude Observatory, National Center for Atmospheric Research, Boulder, Colorado. He is also an Adjoint Professor at the University of Colorado. His interests are in archaeoastronomy, the history of astronomy, the astronomy of the American Indians, and the relation of the sun to the climate of the earth.

Michael Feirtag is a writer and editor specializing in scientific and technological subjects. He is a member of the Board of Editors of *Technology Review,* published at MIT.

Sharon Gibbs is a historian of science currently employed by the National Archives. Her research interests and publications have emphasized the topics discussed in her article: early scientific instruments in the Old World and developing astronomy in the New; among her publications is a book, *Greek and Roman Sundials.*

Owen Gingerich is Professor of Astronomy and of the History of Science at Harvard University, and an astrophysicist at the Smithsonian Astrophysical Observatory.

Jerome Y. Lettvin has been a pocket-pusher in the cleaning industry, an electroplater of golf clubs, sub-assistant writer of horror movies in Hollywood, psychologist of sea sickness, electroencephalographer, designer of lie detectors, and nurse for an octopus colony. He became an uncouched psychiatrist in 1951 and has since been teaching experimental epistemology in the Research Simulation Center at MIT.

Philip Morrison is Institute Professor and Professor of Physics at MIT. His scientific research during the past decade has centered around problems of theoretical physics and cosmology.

Harald A. T. Reiche is Professor of Classics and Philosophy at MIT. His research and publications are chiefly in the area of Greek philosophy and science and their interface with mythology.

INDEX